Handbook of Electrical Measurements

Handbook of Electrical Measurements

Contributors

Gema Romero,Jose Ramon Díaz and Carlos Perez et al.

AURIS
Reference

www.aurisreference.com

Handbook of Electrical Measurements

Contributors: Gema Romero,Jose Ramon Díaz and Carlos Perez et al.

Published by Auris Reference Limited

www.aurisreference.com

United Kingdom

Handbook of Electrical Measurements

ISBN: 978-1-78154-916-2

British Library Cataloguing in Publication Data
A CIP record for this book is available from the British Library

Printed in the United Kingdom

Exclusively distributed by CBS Publishers & Distributors Pvt. Ltd.

Sales & Distribution Rights only for India, Pakistan, Bangladesh, Sri Lanka, Nepal and Bhutan.This book is not to be sold outside these territories.

Contents

List of Abbreviations..vii

List of Contributors..ix

Preface..xv

Chapter 1   Evaluation of Commercial Probes for On-Line Electrical Conductivity
Measurements during Goat Gland Milking Process............................ 1

Chapter 2   Transient Response of Organo-Metal-Halide Solar Cells Analyzed
by Time-Resolved Current-Voltage Measurements 29

Chapter 3   Smart Laser Interferometer with Electrically Tunable Lenses for
Flow Velocity Measurements through Disturbing Interfaces 49

Chapter 4   Measuring User Similarity Using Electric Circuit Analysis:
Application to Collaborative Filtering.. 65

Chapter 5   Electrical Characterization of Traps in AlGaN/GaN FAT-HEMT's on
Silicon Substrate by C-V and DLTS Measurements............................ 89

Chapter 6   Reliability Measurement for Mixed Mode Failures of 33/11 Kilovolt
Electric Power Distribution Stations 101

Chapter 7   Extrinsic and Intrinsic Frequency Dispersion of High-k Materials in
Capacitance-Voltage Measurements 119

Chapter 8   Application of HFCT and UHF Sensors in On-Line Partial
Discharge Measurements for Insulation Diagnosis of High
Voltage Equipment... 157

Chapter 9   Calculation of Constitutive Parameters from Electric and
Magnetic Field Measurements in an Anisotropic Medium
with a Triaxial Instrument 199

Chapter 10   Retinal Vessel Width Measurement at Branchings Using an
Improved Electric Field Theory-Based Graph Approach 221

Chapter 11   Antenna Measurement... 241

Citations... 271

Index.. 273

List of Abbreviations

AO	Adaptive optics
AFM	Atomic force microscopy
ALD	Atomic layer deposition
CET	Capacitance equivalent thickness
CRVE	Central Retinal Vein Equivalent
CF	Collaborative filtering
CMOS	Complementary-metal-oxide-semiconductor
CPC	Compound parabolic concentrating
CUT	Coordinated universal time
CDLTS	Current-deep level transient spectroscopy
DLTS	Deep level transient spectroscopy
DLTS	Deep Level Transient Spectroscopy
ECA	Electric circuit analysis
ELF	Electric line of force
EC	Electrical conductivity
ECA	Electric-circuit analysis
EOT	Equivalent oxide thickness
GM	Galvanometer mirrors
GPS	Global positioning system receiver
HCCP	Hazard and critical control points
HEMT's	High electron mobility transistors
HFCT	High frequency current transformers
HV	High voltage
ITRS	International Technology Roadmap for Semiconductors
KCL	Kirchhoff's current law
KVL	Kirchhoff's voltage law
KWW	Kohlrausch-Williams-Watts
LDV	Laser Doppler velocimeter
MTBF	Mean time between failures
MTTF	Mean time to failure
MTTR	Mean time to repair
MOS	Metal-oxide-semiconductor
MAI	Methylammonium iodide
MHRW	Metropolis–Hastings random walk
MEMS	Micro-electro-mechanic-system
OCT	Optical coherence tomography
OMHP	Organo-metal-halide perovskite
PD	Partial discharge

PCA	Principal component analysis
PPS	Pulse per second signal
RFCT	Radio frequency current transducer
RMS	Root Mean Square
SNR	Signal to noise ratio
SNR	Signal-to-noise ratio
TC	Tanimoto coefficient
TBFs	Time between failures
TL	Tunable lenses
WT	Wavelet transform
XRD	X-ray diffraction

List of Contributors

Gema Romero
Department of Tecnología Agroalimentaria, Escuela Politécnica Superior de Orihuela and Department of Ingeniería de Sistemas y Automatica, nBio Group, Quorum V Building, Campus de Elche, Universidad Miguel Hernández, Spain

Jose Ramon Díaz
Department of Tecnología Agroalimentaria, Escuela Politécnica Superior de Orihuela and Department of Ingeniería de Sistemas y Automatica, nBio Group, Quorum V Building, Campus de Elche, Universidad Miguel Hernández, Spain

Jose Maria Sabater
Department of Tecnología Agroalimentaria, Escuela Politécnica Superior de Orihuela and Department of Ingeniería de Sistemas y Automatica, nBio Group, Quorum V Building, Campus de Elche, Universidad Miguel Hernández, Spain

Carlos Perez
Department of Tecnología Agroalimentaria, Escuela Politécnica Superior de Orihuela and Department of Ingeniería de Sistemas y Automatica, nBio Group, Quorum V Building, Campus de Elche, Universidad Miguel Hernández, Spain

M. Greyson Christoforo
Department of Electrical Engineering, Stanford University, Stanford, CA 94305, USA

Eric T. Hoke
Department of Materials Science an Engineering, Stanford University, Stanford, CA 94305, USA

Michael D. McGehee
Department of Materials Science an Engineering, Stanford University, Stanford, CA 94305, USA

Eva L. Unger
Department of Chemistry, Lund University, 22241 Lund, Sweden

Jürgen W. Czarske
TUD Laboratory for Measurement and Testing Techniques, TUD Faculty of Electrical and Computer Engineering, TUD School of Engineering, TU Dresden, Helmholtzstraße 18, 01069 Dresden, Germany

Hannes Radner
TUD Laboratory for Measurement and Testing Techniques, TUD Faculty of Electrical and Computer Engineering, TUD School of Engineering, TU Dresden, Helmholtzstraße 18, 01069 Dresden, Germany

Christoph Leithold
TUD Laboratory for Measurement and Testing Techniques, TUD Faculty of Electrical and Computer Engineering, TUD School of Engineering, TU Dresden, Helmholtzstraße 18, 01069 Dresden, Germany

Lars Büttner
TUD Laboratory for Measurement and Testing Techniques, TUD Faculty of Electrical and Computer Engineering, TUD School of Engineering, TU Dresden, Helmholtzstraße 18, 01069 Dresden, Germany

Joonhyuk Yang
Graduate School of Culture Technology, Korea Advanced Institute of Science and Technology, Daejeon, Republic of Korea

Jinwook Kim
Department of Electrical Engineering, Pohang University of Science and Technology, Pohang, Republic of Korea

Wonjoon Kim
Department of Management Science/Graduate School of Culture Technology, Korea Advanced Institute of Science and Technology, Daejeon, Republic of Korea

Young Hwan Kim
Department of Electrical Engineering, Pohang University of Science and Technology, Pohang, Republic of Korea

Manel Charfeddine
Laboratoire des Micro-Optoélectroniques et Nanostructures, Université de Monastir, Faculté des Sciences de Monastir, Monastir, Tunisie

Malek Gassoumi
Laboratoire des Micro-Optoélectroniques et Nanostructures, Université de Monastir, Faculté des Sciences de Monastir, Monastir, Tunisie

Hana Mosbahi
Laboratoire des Micro-Optoélectroniques et Nanostructures, Université de Monastir, Faculté des Sciences de Monastir, Monastir, Tunisie

Christophe Gaquiére
Institut d'Electronique de Microélectronique et de Nanotechnologie IMEN, Département Hyperfréquences et Semiconducteurs, Université des Sciences et Technologies de Lille, Villeneuve d'Ascq Cedex, France

Mohamed Ali Zaidi
Laboratoire des Micro-Optoélectroniques et Nanostructures, Université de Monastir, Faculté des Sciences de Monastir, Monastir, Tunisie

Hassen Maaref
Laboratoire des Micro-Optoélectroniques et Nanostructures, Université de Monastir, Faculté des Sciences de Monastir, Monastir, Tunisie

Faris M. Alwan
School of Mathematical Sciences, University Sains Malaysia, Penang, Malaysia
Statistics Department, College of Administration and Economics, Baghdad University, Baghdad, Iraq

Geehan S. Hassan
School of Computer Sciences, University Sains Malaysia, Penang, Malaysia
Ibn-Rushud College of Education, Baghdad University, Baghdad, Iraq

Adam Baharum
School of Mathematical Sciences, University Sains Malaysia, Penang, Malaysia

J. Tao
Department of Microelectronics, Xi'an Jiaotong University, Xi'an 710016, China

C.Z. Zhao
Department of Microelectronics, Xi'an Jiaotong University, Xi'an 710016, China
Department of Electrical and Electronic Engineering, Xi'an Jiaotong-Liverpool University, Suzhou 215123, China
Department of Electrical Engineering and Electronics, University of Liverpool, Liverpool L69 3GJ, UK

C. Zhao
Department of Electrical and Electronic Engineering, Xi'an Jiaotong-Liverpool University, Suzhou 215123, China

Department of Electrical Engineering and Electronics, University of Liverpool, Liverpool L69 3GJ, UK

P. Taechakumput
Department of Electrical Engineering and Electronics, University of Liverpool, Liverpool L69 3GJ, UK

M. Werner
Department of Electrical Engineering and Electronics, University of Liverpool, Liverpool L69 3GJ, UK
Department of Materials Science and Engineering, University of Liverpool, Liverpool L69 3GH, UK

S. Taylor
Department of Electrical Engineering and Electronics, University of Liverpool, Liverpool L69 3GJ, UK

P. R. Chalker
Department of Materials Science and Engineering, University of Liverpool, Liverpool L69 3GH, UK

Fernando Álvarez
Department of Electrical Engineering, Polytechnic University of Madrid, Ronda de Valencia 3, Madrid 28012, Spain

Fernando Garnacho
LCOE-FFII, Eric Kandel 1, Getafe 28906, Spain

Javier Ortego
DIAEL, Peñuelas 38, Madrid 28005, Spain

Miguel Ángel Sánchez-Urán
Department of Electrical Engineering, Polytechnic University of Madrid, Ronda de Valencia 3, Madrid 28012, Spain

Ertan Pekşen
Department of Geophysical Engineering, Kocaeli University, Kocaeli, Turkey

Xiayu Xu
Department of Biomedical Engineering, University of Iowa, Iowa City, Iowa, United States of America

Joseph M. Reinhardt
Department of Biomedical Engineering, University of Iowa, Iowa City, Iowa, United States of America

Qiao Hu
Department of Electrical and Computer Engineering, University of Iowa, Iowa City, Iowa, United States of America

Benjamin Bakall
Department of Ophthalmology and Visual Science, University of Iowa, Iowa City, Iowa, United States of America

Paul S. Tlucek
Department of Ophthalmology and Visual Science, University of Iowa, Iowa City, Iowa, United States of America

Geir Bertelsen
Department of Community Medicine, University of Tromsø, Tromsø, Norway

Michael D. Abra`moff
Department of Biomedical Engineering, University of Iowa, Iowa City, Iowa, United States of America
Department of Electrical and Computer Engineering, University of Iowa, Iowa City, Iowa, United States of America
Department of Ophthalmology and Visual Science, University of Iowa, Iowa City, Iowa, United States of America
Veteran's Administration Medical Center, Iowa City, Iowa, United States of America

Dominique Picard
Supélec Plateau de Moulon 91192 Gif sur Yvette Cedex France

Bernardo Ricardo
Energy and Building Design, Lund University, Lund, Sweden

Davidsson Henrik
Energy and Building Design, Lund University, Lund, Sweden

Gentile Niko
Energy and Building Design, Lund University, Lund, Sweden

Gomes João
Solarus Sunpower AB, Stockholm, Sweden

Gruffman Christian
Finsun AB, Älvkarleby, Sweden

Chea Luis
Universidade Eduardo Mondlane, Maputo, Mozambique

Mumba Chabu
University of Zambia, Lusaka, Zambia

Karlsson Björn
Division of Energy Engineering, Mälardalen University, Västerås, Sweden

Preface

Electrical measurements are the methods, devices and calculations used to measure electrical quantities. Measurement of electrical quantities may be done to measure electrical parameters of a system. The text *Handbook of Electrical Measurements* covers a wide spectrum of topics pertaining to electrical measurements. The aim of first chapter is to evaluate the performance of three commercial conductimeters to be used during mechanical milking of small ruminant halves, especially Murciano-Granadina goats. Second chapter discusses the transient response of organo-metal-halide solar cells analyzed by time-resolved current-voltage measurements. In third chapter, we present a different approach for an adaptive, smart, interferometric laser Doppler velocimetry (LDV) to overcome the former limitations. In fourth chapter, we propose a new technique of measuring user similarity: applying electric-circuit analysis (ECA) to the user–item matrix. High electron mobility transistors (HEMT's) based on Al-GaN/GaN grown by molecular beam epitaxy on silicon substrates have been investigated in fifth chapter. Sixth chapter presents a methodology for assessing the reliability of 33/11 kilovolt high-power stations based on average time between failures. In seventh chapter, the extrinsic and intrinsic causes of frequency dispersion during C-V or C-f (capacitance-frequency) measurements in high-k thin films have been investigated. Eighth chapter proposes an optimized electromagnetic detection method based on the combined use of wideband PD sensors for measurements performed in the HF and UHF frequency ranges, together with the implementation of powerful processing tools. Calculation of constitutive parameters from electric and magnetic field measurements in an anisotropic medium with a triaxial instrument has been discussed in ninth chapter. In tenth chapter, we propose a retinal vessel width measurement method at branch points based on an improved electric field theory motivated graph approach. Last chapter focuses on antenna measurement.

Chapter 1

EVALUATION OF COMMERCIAL PROBES FOR ON-LINE ELECTRICAL CONDUCTIVITY MEASUREMENTS DURING GOAT GLAND MILKING PROCESS

Gema Romero, Jose Ramon Díaz, Jose Maria Sabater and Carlos Perez

Department of Tecnología Agroalimentaria, Escuela Politécnica Superior de Orihuela and Department of Ingeniería de Sistemas y Automatica, nBio Group, Quorum V Building, Campus de Elche, Universidad Miguel Hernández, Spain

ABSTRACT

The measurement of the milk electrical conductivity (EC) during mechanical milking has been widely studied for mastitis detection on cows because its improving of welfare and animal health, although research about small ruminants is scarce. The aim of this study was to evaluate the performance of three commercial conductimeters to be used during mechanical milking of small ruminant halves, especially Murciano-Granadina goats. The objective of this research was to integrate the probes on the milking unit and to check the suitability of the probe selected. The results presented in this research have guided authors to discard the commercial probes and to establish the requirements of a new probe design that is briefly outlined in the conclusions of this contribution.

INTRODUCTION

Food Hygiene regulations approved in 2004 and 2005 by the UE aim to assure public health protection, as well as animal health and welfare. Good agricultural and farming practices need to be implemented and promoted in the primary sector as a first step for a future analysis implementation of hazard and critical control points (HCCP).

Electrical conductivity (EC) devices during milking may improve mastitis detection and thus improve small ruminant welfare and health. It has been widely studied for mastitis detection on cows, due to the capability of automatization in the milking machine [1–4]. The main advantage is the results can be achieved on-line during milking, with objective measurements and with a relatively lower cost, if compared to other mastitis detection methods (bacteriological analysis, SCC, serological, California Mastitis Test, etc.).

Several publications around the automatic measurement of EC for mastitis detection on cows can be found on specialized literature [5–7]. The different methods for on-line EC measurement differ if the measurement is done for the milk coming from the whole udder or at every gland level. In the first case, probe sensors are allocated in the long milk tube (11 onFigure 1) that is the tube beyond the claw (15 on Figure 1), and in the second case in the short milk tube (tubes between 14 and 15 on Figure 1). In both cases EC data are logged during the milking time, when the milk is following through the tubes. Data is processed on a central computer that uses a model or algorithm to detect if the udder or the gland is candidate to be affected by mastitis.

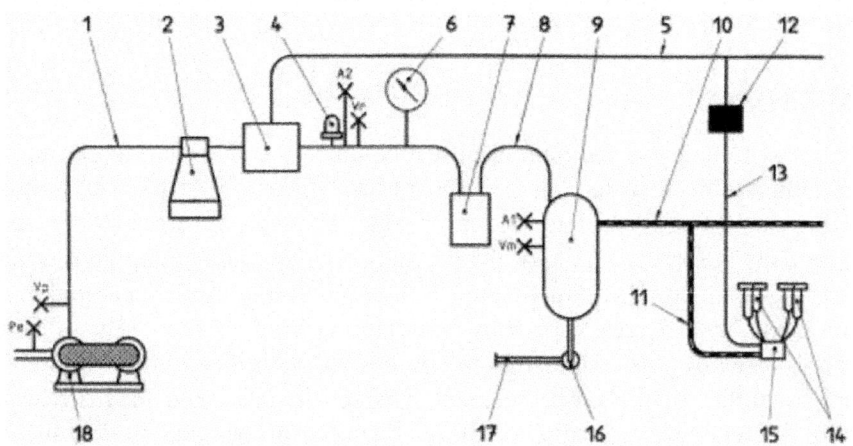

Figure 1: Example of a pipeline milking machine. (extracted from ISO 3819).

Nevertheless, literature about the use of EC measurement for mastitis detection of small ruminants like caprine is very scarce [8–10]. Ying [10] published that mastitis in Saanen breed produces an increase of 0.2 mS/cm (from 5.6 to 5.8 mS/cm) while in Alpina breed the effect was a higher decrease of EC (from 6.1 to 5.4 mS/cm). This previous work invited us to study the different factors related to EC and the association of intramammary infection with EC in the local breed of our region (Murciano-Granadina) [11] obtaining different increases depending on the pathogen affecting. Other aspect was researching on automation of EC measurements during milking by gland, instead of by animal, for mastitis detection of small ruminants.

The aim of this study was to evaluate the performance of three commercial EC probes for their use by gland during small ruminants milking, and to establish the requirements of the desired probe. The goal of this research is to integrate the probes on the milking unit and to check the suitability of the probe selected. The results presented in this research have guided authors to study a new probe design that is briefly outlined on the conclusions of this contribution.

MATERIALS AND METHODS

All the experiments were carried out in the Universidad Miguel Hernández de Elche(UMH, Spain), with the participation of Animal Production and Systems Engineering areas. First experiment was carried out at the laboratory of Systems Engineering, the second one was carried at the laboratory of Animal Production area and the third one was carried out at milking parlour level in the educational and experimental farm of small ruminants of the UMH. Materials and methods of each experiment are detailed below.

Objectives

Prior to trying to develop a suitable system for on-line EC measurement during goat milking, it is necessary to analyze the requirements of the EC probes to be inserted in an on-line system where the intermittent vacuum needed for the massage and the flowing air introduces noise on the measures. Finally, we need to evaluate the performance of commercial probes to suit these requirements.

That is, the objectives covered on this paper are, firstly to study the special requirements of an on-line EC measurement system during milking, specially the effect of fat, vacuum and cleaning process on the temporary performance of the probes, and in second term to evaluate the performance of different commercial equipments for EC measurements, with different physics (inductive, conductive) and geometries in order to know the most desirable properties.

All the tested equipments were acquired to the special tasks of measuring the electrical conductivity of fresh goat milk. The expected result was to be able to define the most suitable physics and geometry of probes for the on-line EC measurement during milking.

Experiments

The study was divided in three experiments, two of them were carried out at laboratory level: off-line experiment and on-line testbed experiment, meanwhile the third experiment was carried out on a real milking parlour, in order to check the performance of the system at field conditions. Depending on the former objectives, next experiments were made in the laboratories of the group.

Off-Line Measurement

The off-line experiment had the aim to select conductimeters with proper measurement to goat milk EC range and to design a prototype to be incorporated to the milking machine in order to take on-line measurements. The measures were done on static process, so the dynamics of the system did not introduce noise.

On-Line Testbed

On this experiment, milking conditions were artificially simulated and a first evaluation of the on-line performance of the commercial probes were done. The geometry of the probes and short milk tubes were studied and a plastic box to accumulate the milk was designed for each probe. The variation on time and cleaning process of the probes were evaluate.

On-Line Milking Parlour

After the second experiment, the selected commercial probes were tested in on-line conditions, at the milking parlour available on the Universidad Miguel Hernandez, during the lactation period of 24 goats. This experiment tested the system on a real milking parlour.

Materials

Next materials were used during experiments:

- Fresh goat milk extracted from Murciano-Granadina goats in a commercial farm.
- UHT cow milk acquired on a supermarket.
- Several saline solutions (NaCl) with different concentrations (different EC).
- Water bath, with regulation of temperature.
- Laboratory EC conductimeter (GLP32 model, Crison Instruments, Allela, Spain), with automatic temperature correction (25 °C or 20 °C). The probe uses conductivity to measure EC, it has two parallel platinum planes (52–92 model from Crison Instruments, Allela, Spain) and a PT1000 temperature probe (model CAT Pt1000 from Crison Instruments, Alella, Spain). This probe was taken as gold reference

to check the measures of the other probes and it was calibrated with standard patrons (1,413 S/cm and 12.88 S/cm) every day (See Figure 2).

Figure 2: Reference conductimeter. Laboratory EC probe.

- Three conductivity meters that were used on the experiments are shown on Figure 3, meanwhile dimensions are summarized on Table 1.

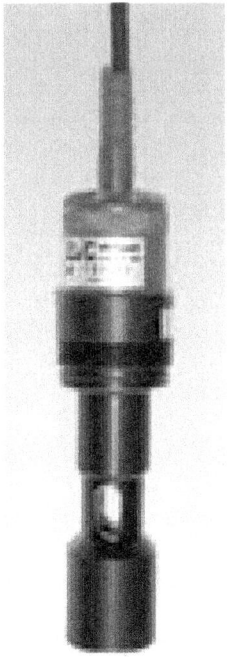

C1

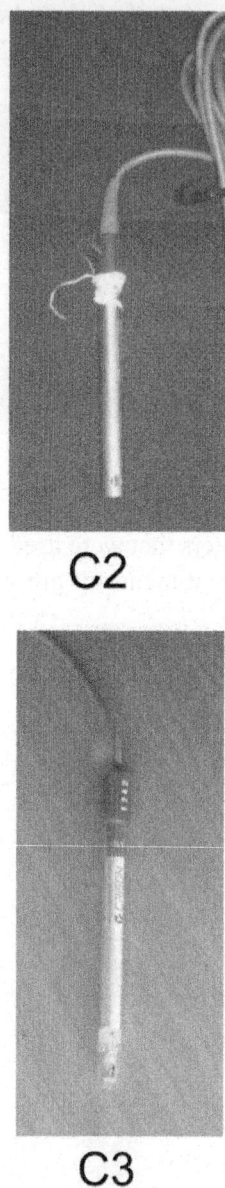

Figure 3: Commercial probes used on experiments.

Table 1: Dimensions of tested probes. Source: own elaboration

Probe	Total length	Diameter	minimum submerged length
C1	94 mm	31 mm	42 mm
C2	130 mm	12 mm	25 mm
C3	130 mm	12 mm	35 mm

- Conductivity meter C1. Model C3655 from B&C Electronics Srl, (Carnate, Milan, Italy), with an inductive probe covered by PVC (Model SI 315 from B&C Electronics Srl, Carnate, Milan, Italy), and an integrated temperature probe PT100. Automatic compensation of temperature is done at 20 °C with a compensation coefficient of 2%/°C. Range of measurement is 0–20 mS/cm. (See Figure 3).

- Conductivity meter C2. Model C3630 from B&C Electronics Srl, (Carnate, Milan, Italy), with an inductive probe made of inox steel (Model 2731312 from B&C Electronics Srl, Carnate, Milan, Italy), and an integrated temperature probe PT100. Automatic compensation of temperature is done at 20 °C with a compensation coefficient of 2%/°C. Range of measurement is 0–20 mS/cm.

- Conductivity meter C3. Model 524 from Crison Instruments, (Allela, Spain), with a conductivity probe made of platinum (Model 52–90 from Crison Instruments, Allela, Spain) and an integrated temperature probe PT100. Automatic compensation of temperature is done at 25 °C with a compensation coefficient of 2%/°C. Range of measurement is 0–199.9 mS/cm.

  a. Portable milking machine (Flaco©, Spain) for goats. This machine was used on the on-line testbed experiment. Milking parameters employed were 40 kPa of vacuum level, 90 puls/min of pulsation rate and 60% of pulsation ratio, usually employed for dairy goats.

  b. Artificial udder made of two halves with a global capacity of 2 liters per gland (see upper right part of Figure 4). A manual valve to regulate the milk flow was allocated on the lower part of each gland. After each valve, a plastic part with nipple shape allows to connect it with the milking machine. This artificial udder was used on the on-line testbed experiment.

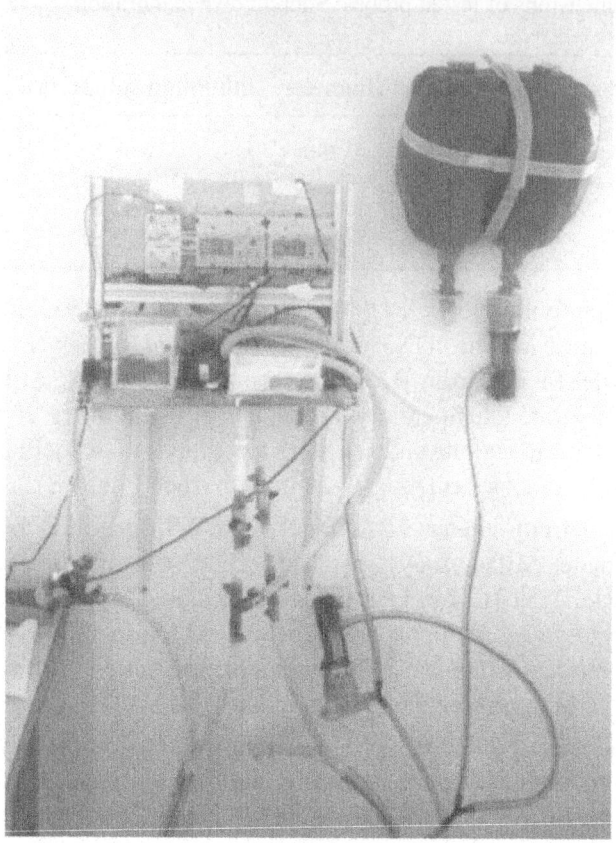

Figure 4: On-line testbed for evaluation of probes.

Each one of the probes were inserted on plastic hand-made chambers designed for each probe (see Figures 4, 5 and 6). These boxes have one input port for the milk and two outports, one for the vacuum and one more for the milk (see Figures 7and 8). This last port has a manual valve that allows to fill the measurement chamber with milk and then, when the valve is opened, milk is evacuated and the measurement process starts again. The sensor probe is allocated on the measurement chamber, and the vacuum flow goes from one of the outports to the input port freely. Dimensions of the boxes varied depending on the dimension of the inner probe. Table 1 reflects the minimum dimension of the measurement chamber that was needed in order to allow the probe to be submerged enough to have a right measurement. These chambers and their dimensions were designed by simple static experiments submerging the probes manually during the off-line experiment.

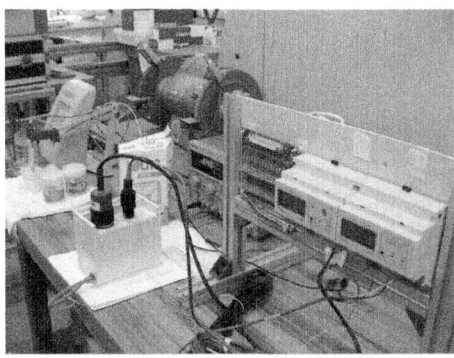

Figure 5: Off-line testbed for evaluation of probes. At rear is the laboratory EC probe and in front are C1 and C2 probes inserted at one measurement chamber.

Figure 6: Detail of the probes in the milking parlour.

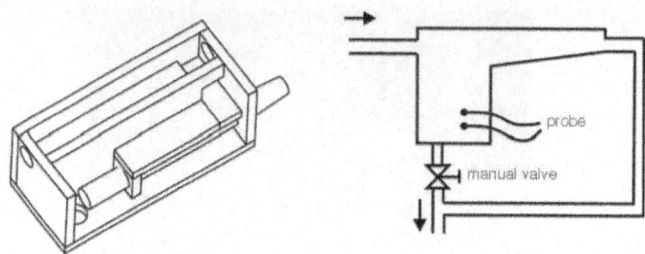

Figure 7: CAD drawing of one of the boxes. Scheme of the inlet and outlet ports.

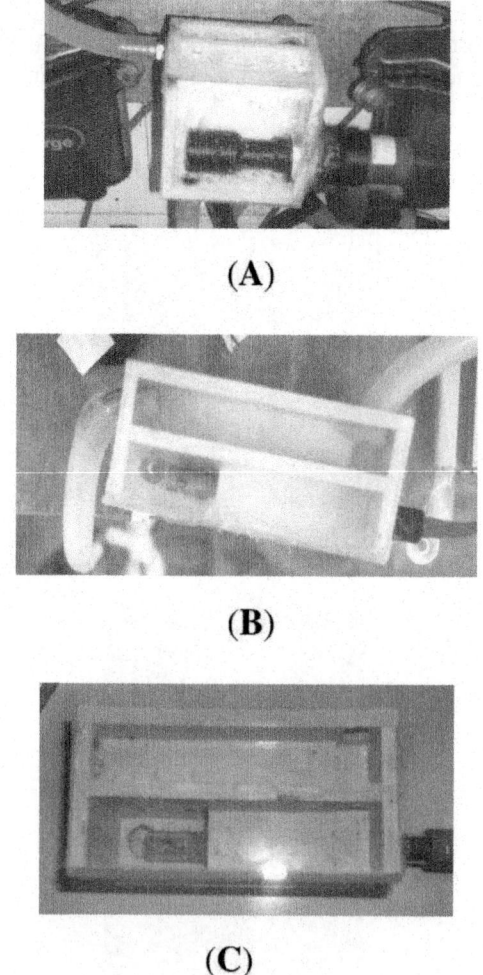

(A)

(B)

(C)

Figure 8: Hand-made boxes for C1 (A), C2 (B) and C3 (C) probes.

Features that were taken account to select the conductivity meters were the range of measurement, the resolution, the temperature compensation and the dynamics of the measure.

- Resolution of the conductivity meter is the minimal unit of EC that can be read; higher the resolution is, lower the value of the minimal unit of EC is. Besides, the resolution depends on the range of measurement, it can be configured depending on the range. Resolution of laboratory probes should be around 1%–2% of the range scale, the standard deviation also being between 1% and 2%.

- Temperature compensation is the ability to modify the measure depending on the temperature of the solution. Temperature variations make change the EC due to the kinetics of fluid particles. Systems with temperature compensation offers the lecture of the EC as it was at normal temperature (usually 20 °C), despite of the real temperature. This compensation is expressed as a percentage, usually 2%/°C for ionic solutions and around 1.5%/°C for acids, alkaline solutions or high concentration solutions.

- Dynamics of the sensor is also an important feature. Some sensor can provide the measure in milliseconds but other uses more time. This is specially important for the temperature probes, as in our experiments, probes are inserted on variable conditions of temperature, from the milk temperature to vacuum conditions.

Other features considered for the probes and conductimeters selection were their functioning principles. Looking at the kind of conductimeters, we can find:

- Conductive probes, which are made of metallic electrodes (platinum or other metal), usually with ring shapes, and work measuring the resistivity of the solution to the electricity transfer from the positive to the negative electrodes. Electrodes are mounted over glass or PVC to avoid deformations.

- Inductive probes, which have the advantage that the fluid does not wet the electrical parts of the probe. Here, two inductively-coupled coils are used. One is the driving coil that produces a magnetic field and it is supplied with accurately-known voltage. The other forms a secondary coil of a transformer. The liquid passing through a channel in the sensor forms one turn in the secondary winding of the transformer. The induced current is the output of the sensor.

- Sanitary probes are inductive probes that have been built using sanitary materials that are not transformed during measurements and do not induces changes on fluid composition.

All the boxes used during experiments (see Figure 8) were hand-made built of PVC and transparent methacrylate for one of the walls so that the user can see the level of the milk on the measurement chamber and manage the manual valve in a proper way.

Methodology

Off-Line Measurement

Figure 5 shows the setup for the off-line measurement of the EC on goat milk. Several boxes with different dimensions were used for each probe to find the minimum measurement-chamber dimension that give us a proper lecture. Different ranges of EC (usual on goat milk) were studied according with the ranges described on [5,10–12].

Several samples, comprising saline water solutions, UHT cow milk and fresh goat milk were measured with the reference conductimeter. Then, samples were manually passed through different volume measurement chambers in order to find the minimum dimension of the needed volume.

On-Line Testbed

First Evaluation of Commercial Probes

A portable milking machine and an artificial udder available at the laboratory (see Figure 4) were used for this experiment. A daily manual milking were done to several goats from a commercial farm. The fresh milk was put on separate jars and translated to the milking laboratory of the teaching farm of the Escuela Politécnica de Orihuela, Universidad Miguel Hernandez. Samples were heated to 37 °C (a similar temperature similar to the inside of the goat udder).

Next, the temperature of each sample was checked, and the EC was obtained using the reference conductimeter, with the same temperature compensation of the checked probe (25 °C or 20 °C). After that, milk was introduced in one of the halves of the artificial udder and it was connected to the milking machine. The emission flow of the udder was regulated to 0.4 L/min. The manual valve allocated after the outlet of the measurement chamber was manually controlled in order to allow the accumulation of milk in the interior of the measurement chamber. When the level of milk was right, i.e., the probe was completely submerged in the fluid, the value of the EC was written down. Next, the valve was opened and the measurement chamber was emptied, closing it afterwards, and starting again to fill the measurement chamber with new milk. The process was repeated until the artificial udder was emptied. Process can be viewed on the scheme of Figure 9.

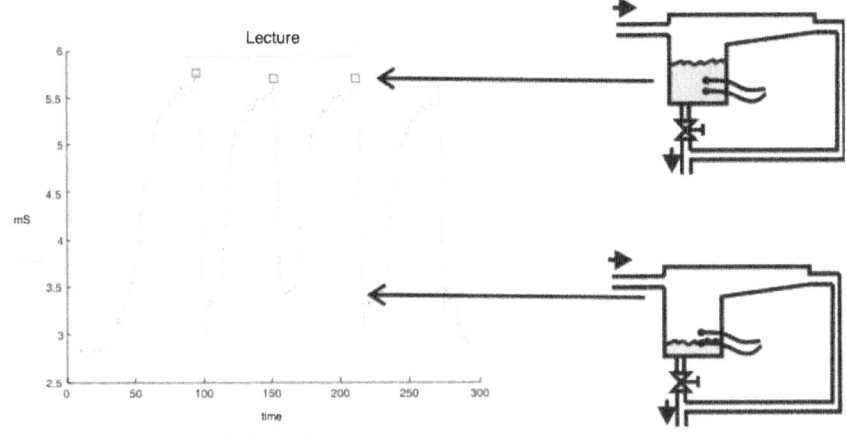

Figure 9: Scheme of the manual valve functioning.

Ten 1 liter sample bottles were milked before cleaning processes, number similar to animals milked during a milking in a commercial farm of small ruminants. The cleaning process included the measurement chamber, the EC probe and the inlet and outlet ports, and it was similar to the conventional cleaning processes on commercial milking machines: open-loop rinsing, closed-loop washing with heated water (40 °C) and alkaline detergent, and again open-loop rinsing. For every 2 washings, a closed-loop washing with acid detergent was added before the last open-loop rinsing in order to avoid deposit of carbonates.

This methodology was repeated for each of the conductimeters 5 times (n = 5 milkings with n = 10 samples each one). On each sample, next data was obtained: deviation between data, difference between the maximum EC and reference one, number of readings with each sample of 1 liter, reading of the appearance of the maximum EC. The average values of each milking for each conductimeter were calculated from the data of the samples.

Evaluation of the Temperature Effect

A second experiment was done using the online laboratory setup. The experiment consisted in evaluating the effect of the temperature variations on the measurement of the probes because in previous experiment a rough effect on the lectures of the probes was observed due to the influence of the vacuum. We postulated that the vacuum generated a great variation of the temperature in the measurement chamber, due to evaporation processes. To check this effect and to study the dynamics of the commercial probes, C2 and C3 conductimeters were used (C1 did not have access to the lecture of the integrated Pt100 sensor). Five 1 liter saline-water samples with different EC were used. Each sample was milked with the same flow and vacuum parameters as milk was before. For each conductimeter, the average and standard deviation of lectures and time of acquisition and order of stable lecture were written down.

On-Line Milking Parlour Experiment

After previous experiments, C1 and C3 conductimeters were selected to be integrated into a real milking parlour. The Educational and Experimental Farm of UMH was selected (see Figures 6 and 10).

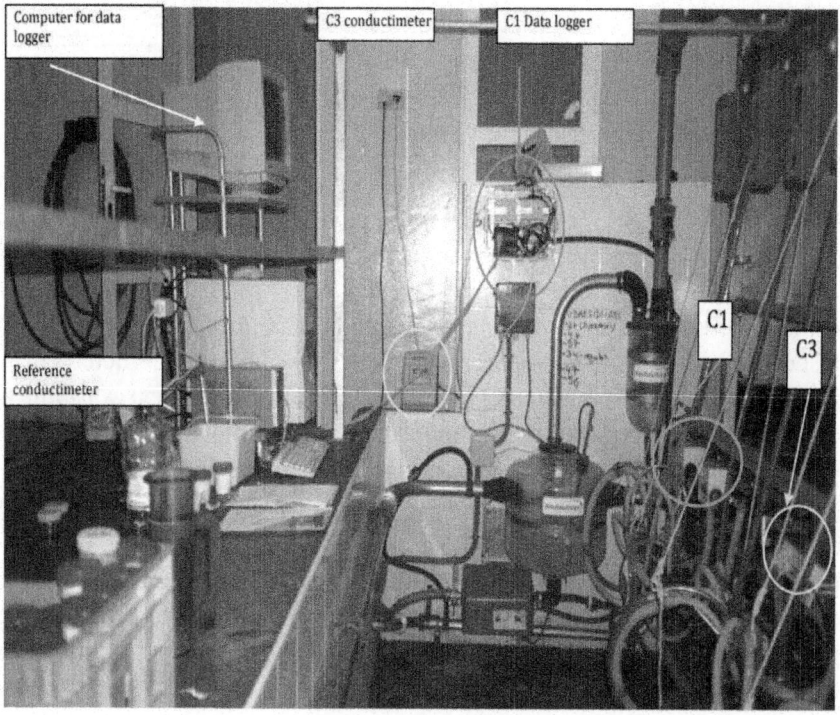

Figure 10: Setup of the conductimeters in the milking parlour.

Twenty four goats at their third month of lactation were selected for this experiment. Animals were homogeneously distributed into 2 groups of 12 goats (24 halves) by group and randomly assigned one conductimeter (C1 or C3) to be tested. Goats were milked daily, and gland milk EC was recorded on-line by the assigned conductimeter, during 120 days at C1 group, and 127 days at C3 group. Tested conductimeters were calibrated at the beginning of the experiment, but never again. In order to check the behaviour of the tested conductimeters, EC of a representative sample of the whole milking of every gland was analyzed using reference conductimeter (CE0) after milking, giving the result with the same compensation of temperature of the tested conductimeter (20 °C if C1 and 25 °C if C3). CE0 was recorded the next sampling days at C1 group: first 45 consecutive days, once after other 15 days, once after other 30 days, and once after other 30 days. At C3 group, CE0 was recorded the next sampling days: first 45 consecutive days, once after 7 days, once after 15 days, once after other 30 days and once after other 30 days. Reference conductimeter was calibrated every sampling day.

Due to the features of the data loggers of each conductimeter, C1 measures were taken with a frequency of 0.5 s, and C3 measures were obtained every 1 s.

C1 data was read directly over Matlab®, using the analog output of the C1 data logger and a 512PG I/O card from National Instruments.

C3 data was read using the RS232 communication port of the C3 data logger. The rough data was stored as a plain text and then imported over Excel® software. Excel® software was used to format the data to a ".csv" file format able to be read from Matlab®.

Using Matlab®, a simple script for getting the EC measures from the rough data was developed. An explanation of the script can be followed on Figure 11. The maximum EC at every fulfill of the measurement chamber was stored and called "local maximum (LM)". The algorithm of the script was similar to the one used in [4] or [3]. The average of the local maxima was calculated for every tested conductimeter. Due to the different volumes of the measurement chamber of C1 and C3 probes, in the case of C1 the average of the three maximum LM was calculated, while for C3 the average of the five maximum LM was calculated. On Figure 11 the rough data and the measures got using the Matlab® script are observed. The effect of the filling of the measurement chamber (explained on Figure 9) can be easily observed. EC2 algorithm was the maximum of the LM calculated.

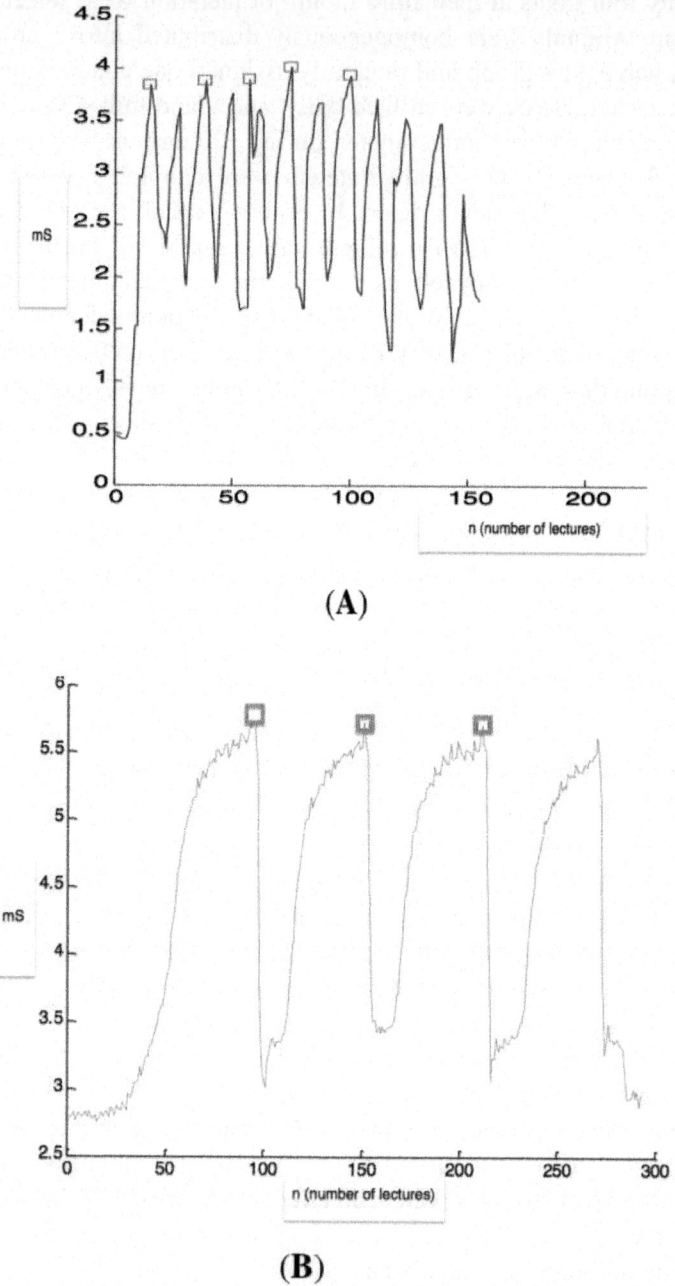

(A)

(B)

Figure 11: Explanation of the selection of the local maxima with Matlab® for the C1 (**A**) and C3 (**B**) probes.

Statistical Analysis

Accuracy of tested conductimeters was analyzed for EC1 and EC2 algorithms considered, comparing with EC0 lecture (EC0-EC1 and EC0-EC2). Average differences and standard deviation of differences was calculated.

The accuracy evolution of tested conductimeters was analyzed among time. Anova analysis of accuracy of EC1 and EC2 algorithms were calculated for both tested conductimeters (Proc GLM, SAS 9.1.). The statistical model used was: $Y_{ij} = \mu + \alpha_i + e_{ij}$. Y_{ij} was the dependent variables (accuracy of CE1 = EC1 — EC0; accuracy of EC2 = EC2 — EC0, of every tested conductimeter: C1 and C3); μ = mean; α_i = the effect of the period of measurement (C1 analysis included 6 levels: the first 45 days of the experiment were divided into 3 levels of 15 days each, 3 levels corresponding to the last measurements in isolated days; C3 analysis included the same first 3 levels and other 4 levels more corresponding to the last measurements carried out in isolated days); e_{ij} = the residual error.

RESULTS AND DISCUSSION

Off-Line Measurement Experiment

The results of this experiment allowed to build the boxes for each one of the probes. Figure 8 shows the boxes developed for C1, C2 and C3. The inner dimensions of the measurement chamber for each probe are detailed on Table 2. Developed boxes were employed in the experiments of on-line testbed and on-line milking parlour.

Table 2: Minimum dimension of the measurement chambers

Probe	Length × Diameter × Submerged Length
C1	80 mm × 50 mm × 45 mm
C2	45 mm × 25 mm × 25 mm
C3	45 mm × 25 mm × 25 mm

On-Line Testbed Experiment

First Evaluation of Commercial Probes

In order to study the evolution of EC among milkings, for every testbed conductimeter we calculated the difference between the maximum EC read by the commercial probe and the reference EC, the number of measurements of a milking, and the order of the lecture where the maximum EC was obtained (see Table 3).

Table 3: Electrical conductivity measurements, deviations from reference, number of lectures during milking, and order of the maximum EC lecture, of the tested conductimeters at on-line testbed experiment

Testbed conducti -meter	Overall EC results				DEC		Number of EC lectures at one milking		Order of lecture of max EC at one milking	
	Number of EC lectures	Average EC (mS/cm)	Average SD (mS/cm)	Average VC %	Average (mS/cm)	SD (mS/cm)	Average	SD	Average	SD
C1	255	6.07	0.32	5.22	−0.34	0.44	5.58	0.67	1.58	0.82
C2	767	6.32	0.52	8.36	−2.46	0.55	15.34	1.73	2.12	1.94
C3	564	6.17	0.06	1.05	−0.33	0.18	11.75	0.47	2.23	1.58

EC: electrical conductivity; SD: standard deviation, VC: variation coefficient; DEC: reference EC-maximum
EC registered by tested conductimeters at every sample; 50 samples were milked by every tested conductimeter;
* The average of the SD (and VC) of lectures obtained at every sample analysis was calculated.

Table 3 shows that the conductimeter which offered more stability and thus a lower average of standard deviation of the measures was the C3 (0.06 mS/cm) following the C1 (0.32 mS/cm), being C2 the probe that presented a larger standard deviation average and thus, a lower repeatability of the readings (0.52 mS/cm). In Figure 12 some examples of the evolution of the measures are shown: C3 presented a more stable lecture of the EC, while C1 and C2 presented some variations due to the parameters of the milking process, as the sample is homogeneous during the milking.

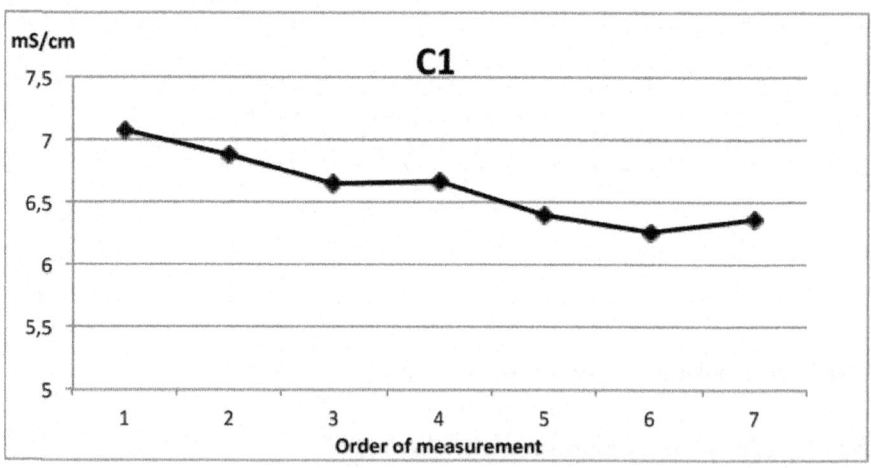

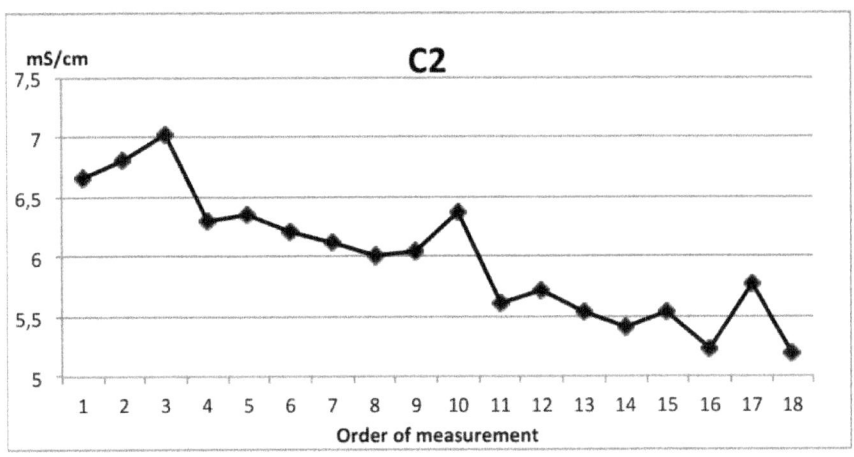

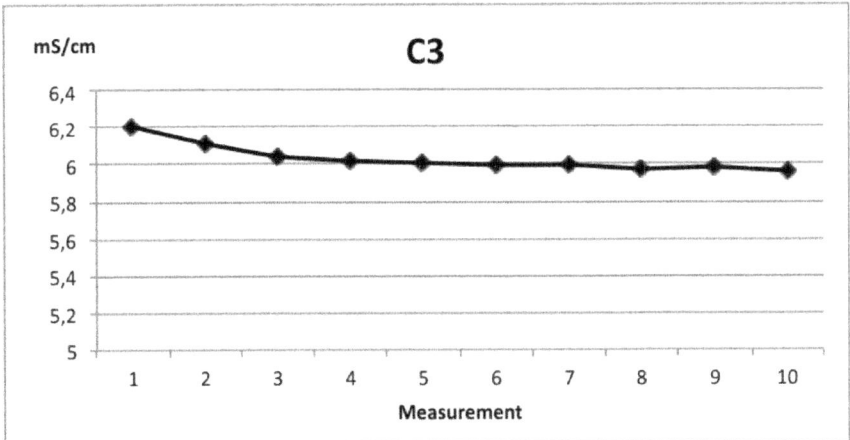

Figure 12: Examples of the evolution of EC measures during one milking.

Regarding to the difference between the maximum values of EC and reference EC, the C3 probe showed the smallest difference in absolute values (0.33 mS/cm), similar to C1 (0.34 mS/cm), and C2 showed larger differences (2.46 mS/cm). This parameter indicates the goodness of the calibration of the probes in comparison with the reference probe. In Figure 13 some examples of the performance of the commercial probes tested with 10 different samples are shown.

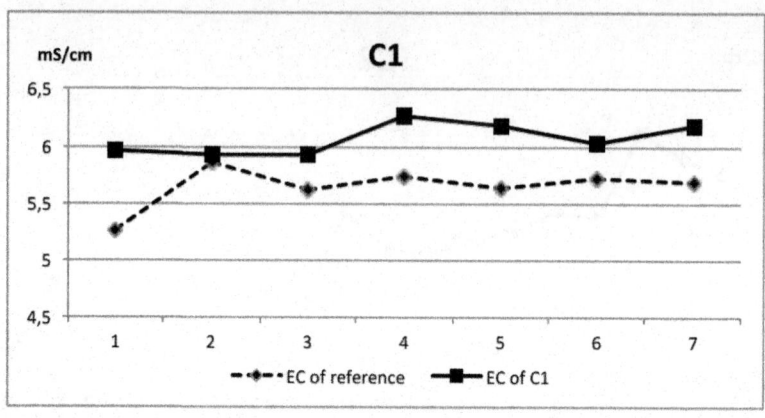

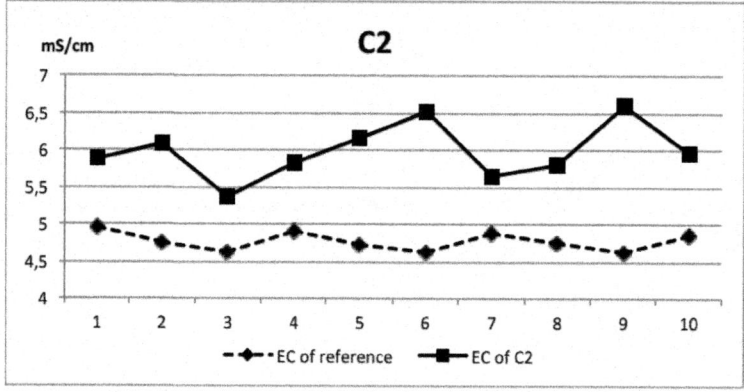

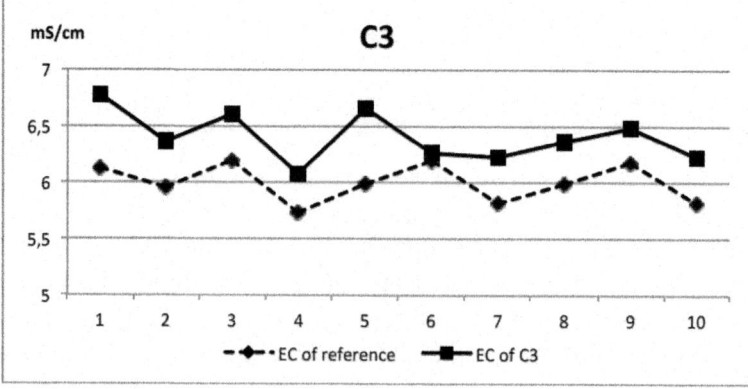

Figure 13: Examples of the performance of commercial probes versus laboratory probe on samples measured consecutively.

Regarding the number of lectures that could be done for 1 liter of sample, C2 achieves a greater number of lectures (15.34) followed by C3 (11.75) and C1 (5.58). This is obvious due to the dimensions of the measurement boxes. It is important due to the nature of the algorithm for the analysis of rough data. As it was shown on Figure 11, the algorithm is dependent of the number of local maxima, which is the number of achieved lectures. So, as much lectures the system is able to do, the better on line measurement of the real EC.

Evaluation of the Temperature Effect

As it can be seen in Table 4, C3 system presented a minor deviation in the temperature lectures (1.33 *vs.* 2.15 °C) together with a faster establishment of the steady state (43 s *vs.* 106 s) which indicates a better dynamic response for working in discontinuous vacuum processes than C1. This feature has a very important influence on the EC measurement, due to the temperature compensation process of the commercial systems. A wrong lecture of temperature will produce a EC error that would avoid its use as a mastitis detection system.

Table 4: Temperature dynamics in C2 and C3 probes at on-line testbed experiment

Probe-sample	Temperature evolution			
	Mean (°C)	Std.dev. (°C)	steady state time (s)	N° of lectures
C2-1	32.53	3.69	120	12
C2-2	32.48	0.93	100	16
C2-3	32.90	1.35	90	16
C2-4	31.20	2.52	100	11
C2-5	31.32	2.27	120	16
Average	32.08	2.15	106	14.2
C3-1	35.38	1.93	28	15
C3-2	34.47	1.45	40	17
C3-3	34.91	1.40	60	17
C3-4	34.61	1.42	62	16
C3-5	31.76	0.44	25	17
Average	34.23	1.33	43	14.4

Regarding the values of Table 5 the same pattern for the evolution of the EC was observed (obvious due to the relation mentioned before). The C2 system had a higher time for steady state than the C3. This could be because of the geometry of the probes. C3 probe is capsuled on crystal while C2 probe on inox steel. The different heat transmission coefficients of materials could be the reason of the differences.

Table 5: EC dynamics in C2 and C3 probes at on-line testbed experiment

Probe-sample	EC evolution			
	Mean (mS/cm)	Std.dev. (mS/cm)	steady state time (s)	N° of lectures
C2-1	9.55	0.32	120	12
C2-2	5.11	0.14	85	16
C2-3	7.13	0.09	90	16
C2-4	5.31	0.20	100	11
C2-5	6.58	0.16	120	16
Average	6.74	0.18	103	14.20
C3-1	5.92	0.06	15	15
C3-2	7.46	0.06	40	17
C3-3	6.24	0.10	20	17
C3-4	8.27	0.07	30	16
C3-5	6.82	0.09	25	17
Average	6.94	0.08	26	14.40

Figure 14 shows an example of the evolution of the EC and temperature during the milking of saline sample on both equipments. It is important to note that the effect of wrong temperature acquisition could have an effect of more than 0.5 mS, which is incompatible with the use of EC for mastitis detection. A probe with less time for steady state is desirable if we want to use it for online mastitis detection.

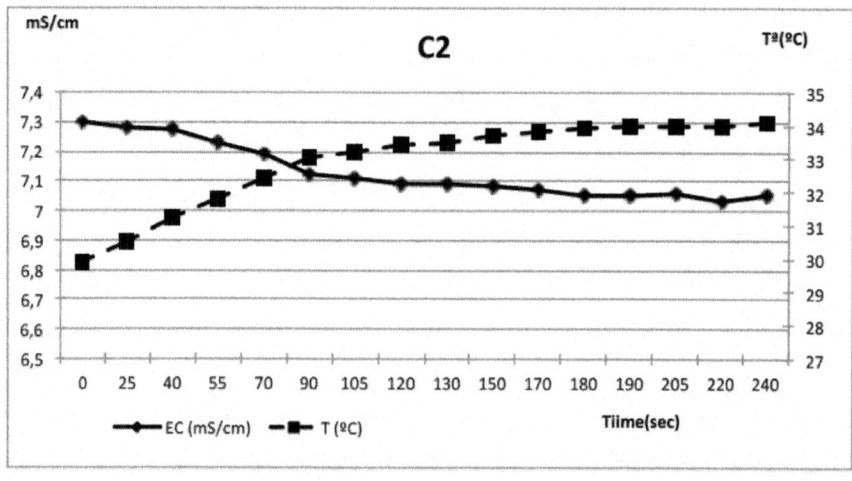

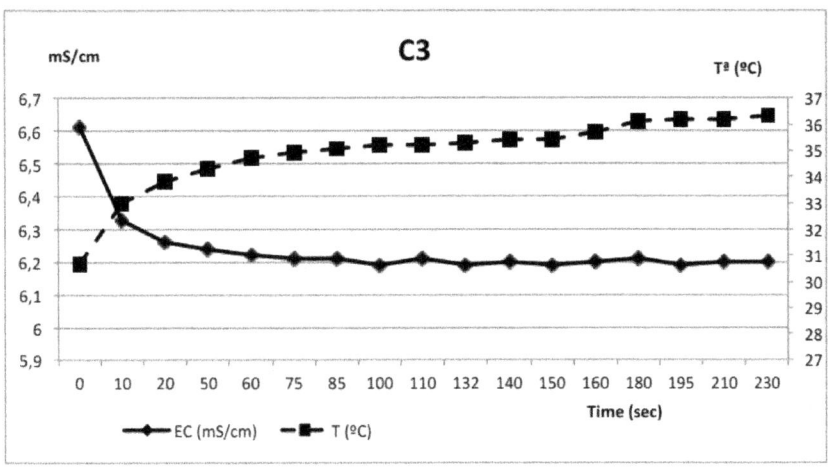

Figure 14: Evolution of lectures of temperature and EC in saline water solutions.

Considering all the results obtained during this testbed experiment, the research team decided to select C1 and C3 for the milking parlour experiment.

On-Line Milking Parlour Experiment

Table 6 shows the results of the first evaluation of C1 and C3 systems in a milking parlour. EC values agree to observed by Park [9], Schuppel and Schwope [8] and Diaz *et al.* [11,12] at different goat breeds.

Table 6: Electrical conductivity (EC, mS/cm) of samples analyzed during on-line milking parlour experiment

	Reference EC				EC1				EC2		
	N	Average	SD	VC	N	Average	SD	VC	N	Average	SD	VC
C1	1,110	5.31	0.46	8.61	1,110	5.62	0.38	6.83	1,110	5.72	0.37	6.52
C3	987	5.73	0.45	7.97	987	6.17	0.59	9.51	987	6.40	0.58	9.01

EC1: See Material and Methods chapter for definition of EC1 and EC2; N: Number of observations considered; SD: standard deviation (mS/cm); VC: variation coefficient (%).

Both developed algorithms to be used by tested conductimeters (EC1 and EC2) presented on-line measures higher than the reference value (CE0). C1 showed values closer to CE0 than C3, indicating a higher accuracy. EC1 algorithm presented a higher accuracy than EC2 for both tested conductimeters. The standard deviation of EC1 (average of local maxima) and EC2 (maximum of local maxima) of C1 were similar and higher than EC0 (reference conductimeter) (0.38, 0.7 and 0.46 mS/cm, respectively). C3 conductimeter presented higher standard deviation than C1 and EC0.

EC1 algorithm had higher accuracy than EC2 for both conductimeters (Table 7) probably because EC1 is the average of several measurements during milking and EC0 was measured in a representative sample of the whole milking, and EC2 was the higher local maxima registered, and obviously higher than EC1. The C1 conductimeter showed lower difference between EC1 and EC2 algorithms (0.1 mS/cm) than C3 (0.23 mS/cm) because C3 registered higher number of measurements at every sample due to the measurement chamber had lower volume, causing a higher dispersion of local maxima values.

Table 7: Accuracy of tested conductimeters during on-line milking parlour experiment

	Accuracy EC1			Accuracy EC2		
	N	Average	SD	N	Average	SD
C1	1,110	−0.31	0.35	1,110	−0.40	0.35
C3	987	−0.44	0.45	987	−0.67	0.38

Accuracy: EC reference (CE0)—EC algorithm tested (EC1 or EC2); N: Number of observations considered; SD: standard deviation (mS/cm).

Both conductimeters suffered an undesirable variation of accuracy among the on-line milking parlour experiment (Tables 8 and 9). C1 reduced the difference with EC0 and, opposite to it C3 increased the difference among time. This uncalibration process questions the proper performance of conductimeters for goats mastitis detection (the application objective) because the expected effect of mastitis on milk EC can be masked by the uncalibration observed.

Table 8: Results of ANOVA analysis of C1 conductimeter accuracy (mS/cm) among time of on-line milking parlour experiment

Accuracy [1]	Statistic	Days of experiment					
		1–15	16–30	31–45	60	90	120
EC1	Mean	−0.425 [a]	−0.361 [b]	−0.225 [c]	0.281 [d]	0.065 [e]	−0.062 [e]
	SE [2]	0.017	0.0175	0.017	0.066	0.068	0.071
EC2	Mean	−0.511 [a]	−0.455 [b]	−0.321 [c]	0.198 [d]	0.060 [e]	−0.229 [e]
	SE [2]	0.017	0.017	0.017	0.066	0.068	0.071
	N	345	346	351	24	23	21

a–e different scripts within a row indicate significant difference ($p < 0.05$); [1] Accuracy: Difference of reference electrical conductivity (EC0) and algorithms considered (EC1 or EC2; See EC1 and EC2 definitions at Material and Methods chapter; [2] SE: standard error; N: Number of observations.

Table 9: Results of ANOVA analysis of C3 conductimeter accuracy (mS/cm) among time of on-line milking parlour experiment

Accuracy [1]	Statistic	Days of experiment						
		1–15	16–30	31–45	60	75	105	135
EC1	Mean	-0.285^{a}	-0.559^{b}	-0.472^{c}	-0.480^{bc}	-0.365^{abc}	-0.408^{abc}	-0.285^{ac}
	SE [2]	0.025	0.025	0.026	0.096	0.097	0.117	0.109
EC2	Mean	-0.493^{a}	-0.793^{b}	-0.714^{c}	-0.772^{bc}	-0.605^{ac}	-0.639^{abc}	-0.675^{abc}
	SE [2]	0.021	0.021	0.020	0.081	0.082	0.099	0.093
	N	302	304	275	21	20	14	16

a–c different scripts within a row indicate significant difference ($p < 0.05$); [1] Accuracy: Difference of reference electrical conductivity (EC0) and algorithms considered (EC1 or EC2); See EC1 and EC2 definitions at Material and Methods chapter; [2] SE: standard error; N: Number of observations.

CONCLUSIONS

An on-line and quick measurement of the temperature is needed to obtain proper EC data for mastitis detection. The special conditions of the milking, with vacuum pulsation and the evaporation due to the vacuum, impose the need of maintaining the probe under immersion continuously, something that was not achieved with the prototypes presented in this paper.

At tested experiments, the better global performance was exhibited by the C3 probe, that had less variation of the lectures and a faster stabilization of the measures. The worst performance was exhibited by the C2 probe that suffered a degradation of the lectures during the milking, maybe due to the deposition of fat on the surface of the probe.

C3 and C1 were employed at on-line milking parlour experiment. The study showed that the on-line values obtained from goat halves were similar to the off-line values reported on the literature. Nevertheless, the dynamics of the lectures of C1 (slow answer to temperature variations) avoid its use of this prototypes as a solely sensor for mastitis detection. Other consideration is about the dimension of the probe. The current dimensions are too big to the expected milk production of a single goat gland, and it reduces the number of lectures taken during the milking. The high correlation with the computation criteria selected (EC1 or EC2) in both probes indicated that trying to maximize the number of lectures is also a requirement (minimize the size) for a better probe design.

So, although the C3 probe presented the best performance to be installed in a milking parlour, it did not give the needed performance to be used as a solely method for on-line mastitis detection. Nevertheless, this study allowed authors to find the requirements for a new probe that minimize the detected problems. This constraints are briefly presented.

Requirements for a New Probe

A new probe design is needed for its use in milking parlours of small ruminants. The design constraints are:

- Small size. Minimizing the size we will be able to have as much number of lectures as possible of every sample.

- Measurement range. The values for Murciano-Granadina breed varies between 4.75 and 6.6 mS/cm, although these values can be elevated by the presence of mastitis.

- To integrate a temperature sensor with fast dynamics.

- The probe must be under immersion continuously. This constraint implies that the measurement chamber should be always full of milk, although this milk should also be recirculated in order not to measure the same sample all the time.

- Auto-calibration of the conductivity meter is needed. The several factors that affect the measurement of the EC and the small differences between a healthy gland and infected halves impose that the conductivity meter should be easy and frequently calibrated.

A conductimeter with this requirements has been presented by the authors under the patent ES U201100694.

ACKNOWLEDGMENTS

The study was supported by project AGL2006-06909 (Ministerio de Educación y Ciencia of Spain and FEDER).

REFERENCES

1. Maatje, K.; de Mol, R.M.; Rossing, W. Cow status monitoring (health and Oestrus) using detection sensors. *Comput. Electron. Agric.* 1997, *16*, 245–254.

2. de Mol, R.M.; Keen, A.; Kroeze, G.H.; Achten, J.M.F.H. Description of a detection model for oestrus and diseases in dairy cattle based on time series analysis combined with a Kalman filter. *Comput. Electron. Agric.* 1999, *22*, 171–185.

3. Mele, M.; Secchiari, P.; Serra, A.; Ferruzzi, G.; Paoletti, F.; Biagioni, M. Application of the tracking signal method to the monitoring of udder health and Oestrus in dairy cows. *Livest. Prod. Sci.* 2001, *72*, 279–284.

4. Zecconi, A.; Piccinini, R.; Giovannini, G.; Casinari, G.; Panzeri, R. Clinical mastitis detection by on-line measurements of milk yield, electrical conductivity and milking duration in commercial dairy farms. *Milchwissenschaft* 2004, *59*, 240–244.

5. Nielen, M.; Deluyker, H.; Schukken, Y.H.; Brand, A. Electrical conductivity of milk: Measurement, modifiers, and meta analysis of mastitis detection performance. *J. Dairy Sci.* 1992, *75*, 606–614.

6. Nielen, M.; Schukken, Y.H.; Brand, A.; Haring, S.; Ferwerda-Van Zonneveld, R.T. Comparison of analysis techniques for on-line detection of clinical mastitis. *J. Dairy Sci.* 1995, *78*, 1050–1061.

7. Norberg, E.; Hogeveen, H.; Korsgaard, I.R.; Friggens, N.C.; Sloth, K.H.M.N.; Løvendahl, P. Electrical conductivity of milk: Ability to predict mastitis status. *J. Dairy Sci.* 2004, *87*, 1099–1107.

8. Schuppel, H.; Schwope, M. Zur Anwendung des Mastitis-Schnelltests und zur messug der elektrischen leitfähigkeit für die kontrolle der eutergesundheit bei ziegen. *Arch. Lebensm* 1998.

9. Park, Y.W. Interrelationships between somatic cell counts, electrical conductivity, bacteria counts, percent fat and protein in goat milk. *Small Rumin. Res.* 1991, *5*, 367–375.

10. Ying, C.; Yang, C.B.; Hsu, J.T. Relationship of somatic cell count, physical, chemical and enzymatic properties to the bacterial standard plate count in different breeds of dairy goats. *Asian-Australas. J. Anim. Sci.* 2004, *17*, 554–559.

11. Diaz, J.R.; Romero, G.; Muelas, R.; Sendra, E.; Pantoja, J.C.F.; Paredes, C. Analysis of the influence of variation factors of electrical conductivity of milk in murciano-granadina goats. *J. Dairy Sci.* 2011, *94*, 3885–3894.

12. Diaz, J.R.; Romero, G.; Alejandro, M.; Muelas, R.; Peris, C. Effect of intramammary infection on milk electrical conductivity in murciano-granadina goats. *J. Dairy Sci.* 2011.

Chapter 2

TRANSIENT RESPONSE OF ORGANO-METAL-HALIDE SOLAR CELLS ANALYZED BY TIME-RESOLVED CURRENT-VOLTAGE MEASUREMENTS

M. Greyson Christoforo [1], Eric T. Hoke [2], Michael D. McGehee [2] and Eva L. Unger [3]

[1]Department of Electrical Engineering, Stanford University, Stanford, CA 94305, USA

[2]Department of Materials Science an Engineering, Stanford University, Stanford, CA 94305, USA

[3]Department of Chemistry, Lund University, 22241 Lund, Sweden

ABSTRACT

The determination of the power conversion efficiency of solar cells based on organo-metal-halides is subject to an ongoing debate. As solar cell devices may exhibit very slow transient response, current-voltage scans in different directions may not be congruent, which is an effect often referred to as hysteresis. We here discuss time-resolved current-voltage measurements as a means to evaluate appropriate delay times (voltage settling times) to be used in current-voltage measurements of solar cells. Furthermore, this method allows the analysis of transient current response to extract time constants that can be used to compare characteristic differences between devices of varying architecture types, selective contacts and changes in devices due to storage or degradation conditions.

INTRODUCTION

The reported device efficiencies of solar cells based on the organo-metal-halide methylammonium lead iodide have increased at an unprecedented pace during the past three years. Impressive performance improvements have been achieved in different device architectures [1,2,3,4] by refinement of the deposition methods of the organo-metal-halide perovskite (OMHP) absorber [4,5,6,7,8] and optimization of selective contact layers [9,10,11]. Determining the steady state power conversion efficiency of solar cell devices based on these novel absorber materials reliably has been to be non-trivial [12,13,14,15].

When the voltage applied across these devices is abruptly changed, the current through the device is often slow to respond, i.e., the corresponding change in current lags behind the change in voltage. These transients in the response can occur on different time scales and often manifest themselves as discrepancies between the forward and reverse current-voltage scans of solar cells, an embodiment which the community commonly refers to as "hysteresis" [12,13,16,17,18,19]. If not taken into account, this "hysteretic" behavior can result in significant over- or underestimation of the solar cell's power conversion efficiency. The absence of hysteresis, i.e., a negligible discrepancy between current-voltage curves when the devices are scanned in the forward and reverse directions, does not guarantee that measurements have been performed correctly. Congruency between forward and reverse scans has been shown both for very short delay times and very long delay times during the current-voltage scan. The delay time or voltage settling time refers to the time delay after a voltage step before the current is sampled [12]. This has caused debate about how to determine the power conversion efficiencies of solar cell device based on organo-metal-halide materials reliably. Slow transient effects in the current-voltage response of solar cell devices are not unique to OMHP-based devices and have also been observed for dye-sensitized solar cells [20]. Recommended measuring protocols have been outlined elsewhere [12,13,21].

Meanwhile, several phenomenological observations on the occurrence or absence of "hysteresis" are important indications of underlying physical phenomena that cause these transient effects. Planar hetero-junction thin film devices are often found to exhibit a more severe hysteresis compared to devices based on meso-porous titanium dioxide [12]. In the latter, thicker meso-porous TiO_2 layers were shown to further reduce hysteresis [19]. In samples with an insulating Al_2O_3 network, on the other hand, severe hysteresis is observed [16,22]. In planar hetero-junction devices, a larger average crystallite size of the organo-metal-halide domains [19], as well as the choice of selective contacts [23,24,25,26,27,28,29] were found to reduce or eliminate hysteresis all together and yield a stable high power conversion efficiency. Often, the only criteria that is analyzed is the presence or absence of "hysteresis", i.e., the congruency between IV-scans in different scan directions, at one single delay time. This is however not sufficient, as OMHP based devices can sometimes be measured "hysteresis-free" or "hysteresis-less" in different time domains, i.e., at different delay times [12]. Devices exhibiting no hysteresis at room temperature conditions may exhibit hysteresis at lower temperature, when processes causing transients slow down [14]. This illustrates, that transient phenomena need to be thoroughly investigated and commented on when stating device performance metrics derived from current-voltage measurements.

The physical origin of the hysteresis in perovskite solar cells is currently under investigation. Research is ongoing to understand the role of charge carrier accumulation in trap states causing chemical capacitive effects [16,30,31,32], the possibility of a ferroelectric polarizability due to re-orientation of methylammonia cations within the crystal cage [32,33,34,35,36,37,38,39,40], and ion migration causing a change in electric field distribution within the active layer and self-doping effects [22,41,42,43]. While capacitive effects due to the trapping and de-trapping of charge carriers are found to occur [30,44], the magnitude of the transient current and voltage response is too large to be caused by chemical capacitive effects alone [12,30]. Various recent contributions have illuminated, that the extraction of photogenerated charge carriers seems indeed more efficient after devices are poled/polarized [13,22,30] at forward bias, for which the more generic term: temporary enhanced by bias (TEBBed) has been suggested [30]. Migration of ionic species has been proven to occur in methylammonium-lead-iodide (MAPI) and related materials, whereof halide vacancies are the most likely to form and migrate [22,43,45,46,47,48]. The intrinsic defect density and chemistry is critically influenced by the materials deposition and processing conditions [49,50,51,52,53]. Experimental evidence corroborates that the re-distribution of charged defects under an applied electric field causes a change in the internal potential gradient, resulting in a temporary p-i-n or p-n homo-junction in the MAPI layer that explains the temporarily enhanced charge carrier extraction efficiency [22,30,42,52].

The observed transients in the current-voltage response are very likely related to temporal and bias-dependent changes in the distribution of charged defects in the absorber layer. Systematic methods to quantify and distinguish between transient effects are needed to further our understanding of the electronic and chemical processes in solar cell devices based on OMHPs. One metric used to quantify the difference between the forward and reverse IV-scan is the hysteresis index, i.e., the difference in the current generated at 80% of the V_{oc} between the reverse and forward scan direction divided by the current measured at this voltage in the reverse scan direction [19]. However, this metric is not well-defined since the "hysteresis index" is typically highly dependent on the voltage scan rate: for example negligible discrepancy between the forward and reverse scan direction can often be obtained at both extremely short (ms) and very long (>1 s) delay times while pronounced "hysteresis" is observed at intermediate times [12]. One of the most reliable ways to compare device performance is tracking their steady-state power conversion efficiency over time [12,16].

Time-resolved staircase voltammetry measurements are a valuable mean to monitor the current response during a voltage step (Figure 1) and from this evaluate appropriate delay times for current-voltage measurements to ensure that measurements are carried out under steady-state conditions [12]. This profile of consecutive applied potentials can also be referred to as a (linear) staircase sweep [53] and is commonly employed to evade the contribution of capacitive charging events when analyzing electrochemical systems. We, herein, highlight the unique picture and insight into a solar cell's dynamic response that can be gained by continuously recording the current through the cell during a voltage staircase sweep, that we will be referring to as I,V(t)-measurements. This method is in contrast to the typical I(V)-measurement that only records one value of current for each step in voltage. I,V(t) measurements are particularly useful to quantify "memory" or "hysteresis" effects in solar cells based on OMHPs and allow the determination of the minimum delay time to reach steady-state during a regular I(V)-measurements as visualized in Figure 1. Since the current is being continuously recorded, we can in principle derive the resulting I(V)-curve for any delay time through data processing. Furthermore, the time constants of the current transients in each segment can be extracted and compared as a function of voltage. Additionally, integrated currents in each voltage segment can be compared to evaluate changes in the internal electric field of the devices caused by light-soaking and/or applied potentials (TEBBing [31]).

To perform the measurements described herein, we wrote a measurement routine to control a Keithley 2400 sourcemeter in the Python programming language [54]. A MATLAB script was used to extract and analyze the I,V(t)-data generated [55]. Both scripts are freely available on the web. We hope this code will be useful to other researchers in the community and we welcome additions and improvements to it.

EXPERIMENTAL SECTION

The device data shown herein was acquired on solution-processed thin film solar cell devices prepared and tested within the scope of our previous report and devices were fabricated as described therein [12] In short, the n-type selective contact was a titanium dioxide layer, prepared by spray-pyrolysis deposition of a diluted solution of Titanium-di(isopropoxid)-bis(acetylacetonate) (Aldrich) onto a fluorine-doped tin oxide glass substrate (Pilkington, TEC15). Methylammonium-lead-iodide (MAPbI$_3$) was deposited from a solution containing three equivalents of the methylammonium iodide (MAI) and lead chloride (PbCl$_2$) precursors, with MAPbI$_3$ being formed upon sublimation of methylammonium chloride (MACl) during annealing at 100 °C for 45 min [56]. The p-type contact was established by spin-coating a solution of spiro-OMeTAD, the oxidized analogue spiro(TFSI)$_2$ and the additives lithium-trifluoro and tBP [57]. A gold back contact of 0.2 cm^{-2} was deposited by thermal evaporation. For measurements, solar cells were masked with an aperture of 0.12 cm^{-2} defining the active area. Prior and between measurements, samples were stored in a dry, dark environment. The effect of conditions prior to the measurements will be discussed in the Results and Discussion section.

Current response during staircase voltage sweeps, I,V(t) measurements, were measured under simulated AM1.5 solar irradiation from a solar simulator (Newport). The device area was defined by using a shadow mask with an area similar in size to the active area of the evaporated metal contact. In a typical measurement, the voltage across the cell was stepped from short-circuit to a forward bias of at least 1.1 V (forward scan), or from forward bias to 0 V (reverse scan) in 50 mV increments with a 15 s dwell time during each step and a current and voltage measurement sampling rate of about 140 Hz. In each voltage segment, both for current and voltage, roughly 2100 values are hence recorded. The ultimate output of this measurement contains three data columns: time, voltage and current. Figure 1 shows the I(t) and V(t) (staircase profile) on a shared time axis. Data analysis was carried out with a MATLAB script that allowed us to plot the resulting I(V) curve as a function of delay time (Figure 2d). Furthermore, the current response in each constant voltage segment can be analyzed by fitting to a characteristic equation (using a Non-linear Least Squares method). For the data analysis performed herein, a bi-exponential was found to fit the transient current data reasonably well, although some other equation may well be more appropriate to describe the underlying physical processes causing the transients. Integration of the transient current response allowed us to visualize at which applied potentials the transient response causes the largest discrepancy with respect to the steady state response of the solar cell device.

RESULTS AND DISCUSSION

Time-Resolved Current-Voltage-Measurements

Figure 1 shows typical output data from the I,V(t) measurement. This particular scan was a reverse scan (from forward bias to 0 V) performed in voltage steps of 50 mV and with a dwell time of 15 s after each step. The I,V(t) measurement provides a unique picture of the solar cell's dynamic operating characteristics since it captures the transient decays or rises in current in response to the step changes in voltage. These dynamics are not captured by conventional I(V) measurements. The I(V) plot can be derived by averaging the current over a defined sampling time t_s after waiting a defined delay time t_d (Figure 1, inset A) after each voltage step. The delay gives the current time to settle to a constant value, which avoids distortion of the resulting I(V) data due to capacitive charging or discharging currents. Delay times t_d typically on the order of 10 seconds are required to avoid distortions of the IV-curve due to transient effects commonly observed in perovskite devices.

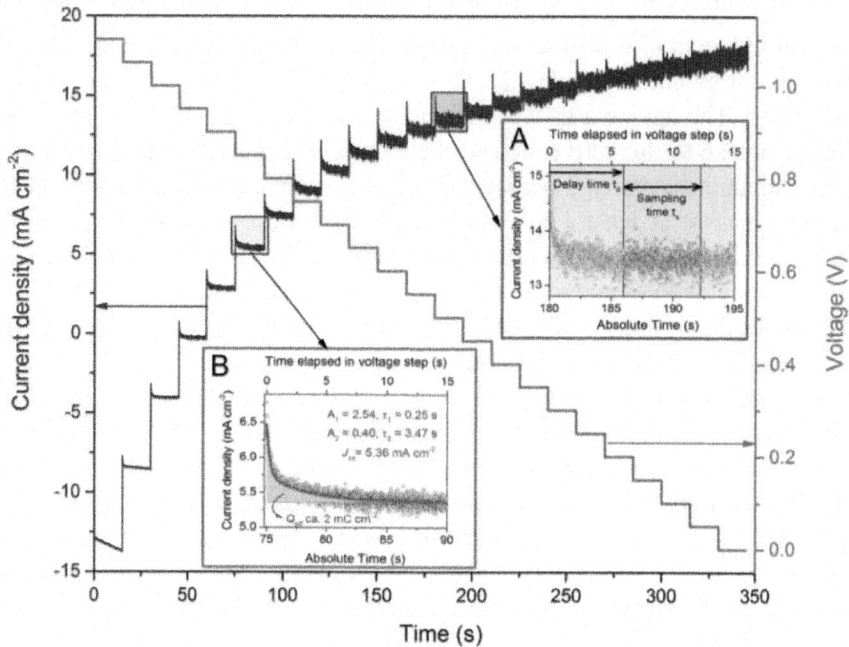

Figure 1: Transient current-density (left axis) response as a function of time (x-axis) in response to a staircase voltage sweep (right axis) from 1.1 V to 0 V (reverse scan direction). The current density exhibits visible overshoots followed by a slow decay in each voltage segment. Inset (A) highlights that both the delay time t_d and sampling time t_s influence the current recorded during a regular I(V) measurement. Inset (B) illustrates the possibility to analyze each individual voltage segment to extract characteristic time constants, (τ_1 and τ_2), pre-exponential factors (A_1 and A_2), the steady state current (J_{ss}) and integrated current density yielding the effective displacement charge density (Q_{eff}).

Familiar I(V) characterization curves can be recreated from the I(V,t) dataset analytically by choosing t_d, and t_s, as illustrated in Figure 1, inset (A). The impact of both t_d and t_s on the resulting I(V) curve can then be evaluated. While seemingly more time consuming and complicated than a simple I(V) sweep, one should consider the wealth of information collected in a single I,V(t) measurement. We found this procedure indeed to be less time consuming than trying to find an optimal scan rate by iteratively increasing the delay time in a regular I(V) measurements until negligible discrepancy between forward and reverse scan direction was reached. Importantly, the I,V(t) measurement gives a clear indication of the time scale and magnitude of transient effects in the solar cell devices.

The data shown here illustrates that for this particular thin film solar cell device based on methylammonium-lead-iodide with an n-selective titania and a p-selective spiro-MeOTAD contact, the response of the photocurrent to a change in applied potential is very slow and a $t_d > 5$ s should be employed. A shorter t_d would have resulted in a substantially higher photocurrent, especially around the maximum power point, for the device scanned in the reverse scan direction.

In Figure 2, we compare the I,V(t) measurements of the same devices in the reverse (a, equal to Figure 1) and forward (b) scan direction. The forward scan exhibits the opposite transient response compared to the reverse scan exhibiting an immediate drop and subsequent exponential rise in the photocurrent in each voltage segment. In Figure 2c, the two measurements were combined into one graph by reversing the time-axis for the reverse scan. We chose to discuss this example as it illustrates, that the steady state current densities, J_{ss}, are not superimposed between 0.4 V and 0.8 V for the reverse and forward scan with the forward scan exhibiting slightly higher values for J_{ss}. While this may be considered a negative example it highlights an important phenomena: The I(V)-characteristics of OMHP based solar cells depend on their measurement history and may change during the measurement.

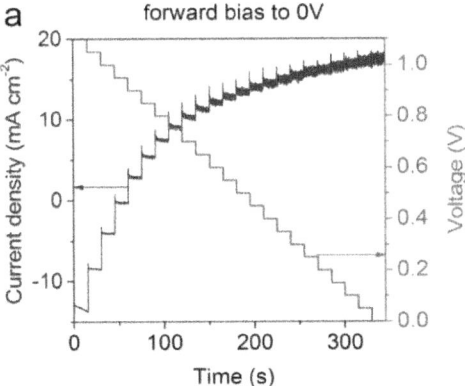

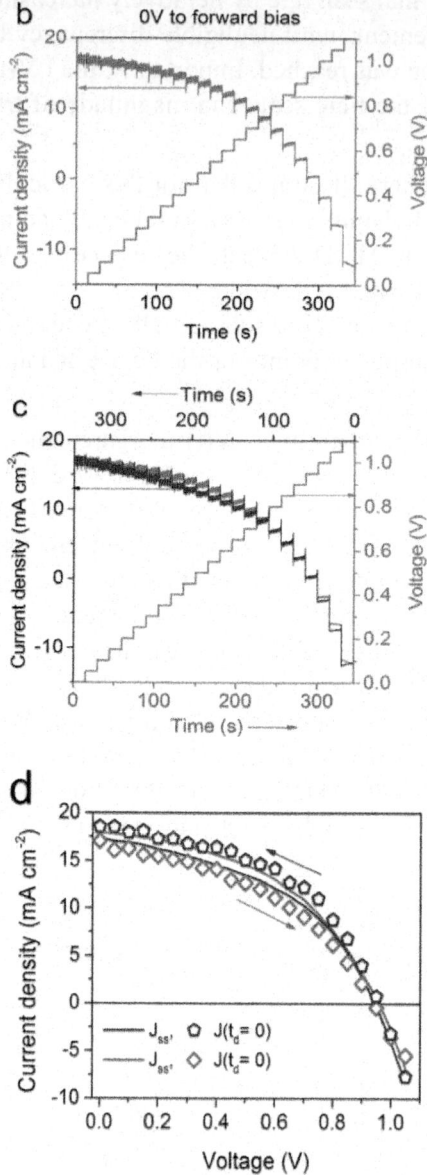

Figure 2: I,V(t) scan in the reverse scan direction (a) exhibits an over-shoot followed by a transient decay while in the forward scan direction (b) the photocurrent drops followed by an exponential increase during each potential step. The overlay of both data set in (c) was created by reversing the time axis for the forward scan. Plot (d) shows the resulting I(V) curves derived from the I,V(t) plots (a–c) using either the steady-state current density, J_{ss}, or the current density at negligible delay time $J(t_d = 0)$ for the current density.

Figure 2d shows I(V) curves generated from the the I,V(t) data both for the the forward and reverse scans shown in plots a and b. The circles correspond to the first measurement point at $t_d = 0$ and $t_s = 0$ while the lines represent the resulting steady-state current density J_{ss} by fitting the current transients to Equation (1). Apparently, the steady state current density J_{ss} does not converge for the forward and reverse scan with the forward scan leading to larger values of J_{ss} compared to the reverse scan. Metrics for the power conversion efficiency may hence vary greatly depending on the measurement conditions, in this case from 6.7% to 8.5%.

To rationalize the higher steady state current density measured in the forward scan we would like to note that the forward scan was in this case performed immediately after the reverse scan. The results are evidently influenced by the measurement history of the device, which may be the result of ion redistribution or other effects leading to changes in the internal field distribution [22,30,42,52] causing differences in the charge carrier extraction efficiency. Additionally, the temperature of the device may have increased during the prolonged measurements under illumination, leading to a higher current density. This example demonstrates that true steady-state conditions are indeed very difficult to achieve and maintain in methylammonium-lead-iodide based solar cells. We found the staircase voltammetry measurements shown here to be instrumental in identifying both the fast, multi-time scale transient dynamics following each voltage segment as well as the slower, transients taking place over the course of minutes that are apparent in discrepancies between consecutive staircase sequence measurements. These results emphasize the importance of performing measurements rigorously, providing detailed information on the measurement routines employed, commenting on the measurement history of a device and, very importantly, also commenting on possible deviations from an ideal or consistent behavior as this may offer important indications on the long term stability of the devices.

In this context, we would also like to re-iterate the often ignored difference between the scan rate and the delay time when specifying the experimental conditions used for I(V) measurements. The scan rate of any I(V) measurement is defined by the voltage step size and dwell time in each voltage step. For the measurement shown in Figure 1 the scan rate is hence 3.33 mV/s (50 mV/15 s), independently of which t_d is used. We therefore recommend stating the delay time and step size rather than scan rate when specifying experimental I(V) measurement conditions. Another remark: there is no optimal or defaultdelay time appropriate for measuring all solar cells. For solar cells based on OMHPs in particular, the minimum t_d to measure devices under quasi-steady-state conditions is strongly architecture and device dependent and should be re-evaluated every time. In this regard, the I,V(t) measurement is both a means to determine the minimum t_d and also capture and analyze the devices transient response, which may provide insight into differences between device types, effects of selective contacts, device polarization (TEBBing), light-soaking and degradation effects among other things.

Analysis of the Transient Response

As indicated in Figure 1, inset (B), the transient current response in each voltage segment can be analyzed further by fitting the experimental data. We found that a bi-exponential fit is fairly adequate to capture the device's transient current response to a step change in voltage:

$$J(t) = J_{ss} + A_1 \exp\left(-\frac{t - t_0}{\tau_1}\right) + A_2 \exp\left(-\frac{t - t_0}{\tau_2}\right)$$

(1)

Here t_0 is the time at the beginning of each voltage segment, J_{ss} is the aforementioned steady-state current, τ_1 and τ_2 are the characteristic time constants, and A_1 and A_2 are exponential scaling factors. For the forward scan, the scaling factors A_1 and A_2 adopt negative signs as the equation describes an exponential rise rather than decay for the scan in forward direction (see Figure 2b). Note that while we've found empirically that a bi-exponential function fits the data quite well, other fit functions or combinations of fit functions may be more suitable to describe the underlying physical phenomena causing the transient current response. As discussed recently in the literature, ion migration is the likely cause for the observed transients as they affect the device polarization [43].

With the bi-exponential fit to fit our experimental data (1), we can extract two time constants that we refer to as τ_{fast} and τ_{slow}, which we attribute to two distinct physical processes with obvious differences in rate. In Figure 3, we compare the time constant for the forward (fwd) and reverse (rev) scan as a function of the applied voltage. The fast τ_{fast} and slow time τ_{slow} constants are of a similar order of magnitude for both scan directions with the slow time constants τ_{slow} being approximately an order magnitude longer compared to τ_{fast}. Both τ_{slow} and τ_{fast} exhibit a dependency on the applied potentials, becoming shorter towards short circuit. In direct comparison, the τ_{slow} here appears to be a factor of 2–4 longer for the forward compared to the reverse scan while τ_{fast} appears to depend much less on the scan direction. Note that the forward scan was carried out after the reverse scan.

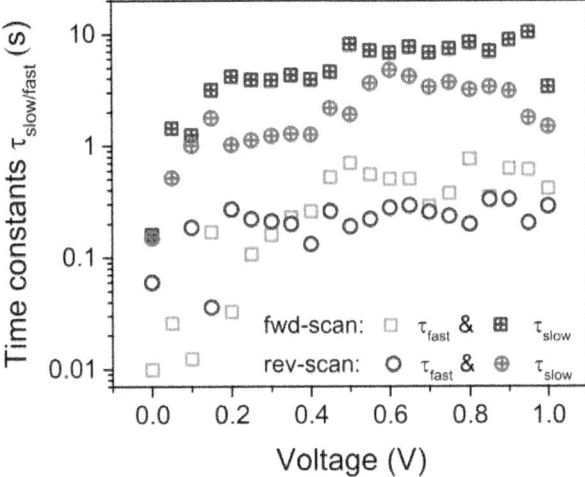

Figure 3: Fast and slow time constants (τ_{fast} and τ_{slow}) extracted by fitting the transient current response in each voltage segment to the bi-exponential fit function given in Equation (1) both for the forward (fwd) scan and for the reverse (rev) scan direction.

The magnitude of τ_{slow} is comparable to the characteristic slow time constants extracted by Ono et al. [53], who performed a similar analysis on OMHP solar cells based on meso-porous titanium dioxide electrodes. They attributed the slow time constant to the change in device polarization due to the ferroelectric properties of the OMHP materials. Ferroelectric polarization due to the re-orientation of the methylammonium cation in the inorganic lattice has been discussed as one possible cause for the hysteresis phenomenon [16,17,39]. However, recent publications on this topic have highlighted that the time scale of molecular re-orientation should be on the pico-second time scale [58] and is hence unlikely to be the cause for the transient phenomena occurring on a second time scale. Dipole reorientation may however affect the band structure and possibly charge carrier dynamics in methyl-ammonium-lead-halides [36,37,59].

Our interpretation of the slow transient is more in line with changes in the internal electric field in the OMHP absorber layer due to electrode polarization caused by redistribution of ionic constituents [13,14,38,46,60]. Migrating halide vacancies in the absorber layer under bias can reasonably cause self-doping effects at the selective contacts transiently forming a p-i-n homo-junction that facilitates charge carrier extraction [22,43,46]. Ionic vacancies are both likely to influence the electric field distribution in the absorber and cause self-doping but could simultaneously also act as charge carrier trapping sites.

The transients in the photocurrent cannot hence be attributed to capacitive charging or discharging currents. Integration of the current transient does not give values representative of capacitive charge carrier densities stored in the device. However, integration of the transient portion of the current response can give insight into the voltages at which the device response is most affected by transient phenomena. In Figure 4, we show the integrated current density from each voltage segment for the reverse (squares) as well as the forward scan (circles) distinguishing between the corresponding integrated current for the fast (open symbols) and slow time constants (plus symbols). This plot illustrates that it is predominantly the slow process that causes the discrepancy in the current-voltage response and that the effect is approximately symmetrical, peaking close to 0.7 V, for the reverse and forward scan. The latter exhibits a negative signal as the transient is negative with respect to the steady state current as shown in Figure 2b. Noticeably, the magnitude of the integrated current is almost twice as large for the forward scan compared to the reverse scan. For the data shown here, the forward scan was carried out after the reverse scan. We find that the magnitude of the integrated current density transient depends on the conditions the devices was exposed to for a significant time prior to the start of the measurement. These results highlight, that devices change dynamically in response to bias or light. Storage and conditions prior to the current-voltage measurements can have a dramatic influence on the transient response of the device as the distribution of ionic species may indeed be quite different.

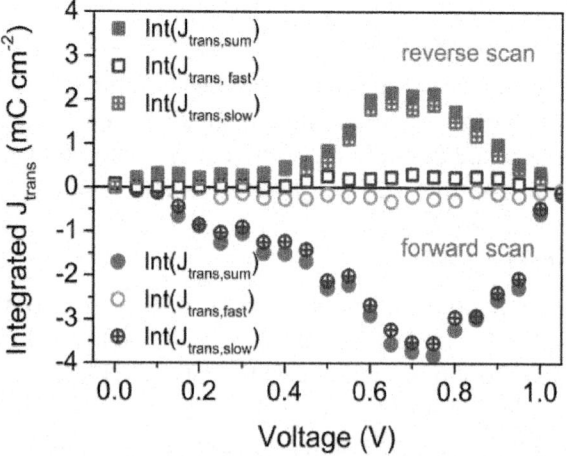

Figure 4. Fast and slow time constants (τ_{fast} and τ_{slow}) extracted by fitting the transient current response in each voltage segment to the bi-exponential fit function given in Equation (1) both for the forward scan and for the reverse scan direction.

CONCLUSIONS

Evidently, it is not trivial to find reliable measurement conditions and protocols to determine the current-voltage response and derive performance metrics for solar cells based on organo-metal-halide semiconductors. We found that the staircase voltammetry measurement routine described herein provides a method for both characterizing and visualizing the transient response of solar cells during I(V) measurements and for determining appropriate measurement delay times that are sufficiently long to produce quasi-steady-state conditions in the device for current-voltage measurements. The data shown here highlights that the time constants of processes causing the slow transient response in solar cells based on $MAPbI_3$ are strongly affected by prior measurements and storage conditions (light, bias, etc.) and that the devices dynamically change during a current-voltage scan. This creates a considerable uncertainty when defining the "stabilized" power conversion efficiency of solar cells based on an I(V) sweep of an OMHP based solar cell. It is important to remember that solar cells are expected to operate at their stated performance specifications for tens of years after installation. Intelligent tracking of the device's maximum power point performed under continuous full illumination for an extended period of time is most certainly the most honest and accurate way to evaluate the performance and stability of these solar cells. Evaluation of the time constants and the magnitude of transient currents during current-voltage measurements as described herein will prove useful in comparing devices of different architecture types and contact layers and can also give insight into changes in devices during degradation or due to previous exposure conditions.

ACKNOWLEDGMENTS

E.L.U. thanks the Marcus and Amalia Wallenberg memorial fund for a postdoctoral research fellowship. Experimental work was funded by the Global Climate and Energy Project (GCEP). M.G.C. has received funding from the European Union's Horizon 2020 research and innovation programme under the Marie Skłodowska-Curie grant agreement No 659667. This work reflects only the author's viewand that the EU is not liable for any use that may be made of theinformation contained therein.

AUTHOR CONTRIBUTIONS

E.L.U. conceived the measurement methodology, performed experiments and data analysis and was the main responsible for writing this manuscript; M.G.C. programmed the python scan routine, wrote the MATLAB script and co-authored this manuscript; E.T.H. helped in the development of the step-IV measurement and co-authored this manuscript; M.D.M. gave input and support to this work.

REFERENCES

1. Lee, M.M.; Teuscher, J.; Miyasaka, T.; Murakami, T.N.; Snaith, H.J. Efficient Hybrid Solar Cells Based on Meso-Superstructured Organometal Halide Perovskites. Science 2012, 3, 1–5.

2. Kim, H.-S.; Lee, C.-R.; Im, J.-H.; Lee, K.-B.; Moehl, T.; Marchioro, A.; Moon, S.-J.; Humphry-Baker, R.; Yum, J.-H.; Moser, J.E.; et al. Lead iodide perovskite sensitized all-solid-state submicron thin film mesoscopic solar cell with efficiency exceeding 9%. Sci. Rep. 2012, 2.

3. Docampo, P.; Ball, J.M.; Darwich, M.; Eperon, G.E.; Snaith, H.J. Efficient organometal trihalide perovskite planar-heterojunction solar cells on flexible polymer substrates. Nat. Commun. 2013, 4.

4. Liu, M.; Johnston, M.B.; Snaith, H.J. Efficient planar heterojunction perovskite solar cells by vapour deposition. Nature2013, 501, 395–398.

5. Burschka, J.; Pellet, N.; Moon, S.-J.; Humphry-Baker, R.; Gao, P.; Nazeeruddin, M.K.; Grätzel, M. Sequential deposition as a route to high-performance perovskite-sensitized solar cells. Nature 2013, 499, 316–319.

6. Chen, Q.; Zhou, H.; Hong, Z.; Luo, S.; Duan, H.-S.; Wang, H.-H.; Liu, Y.; Li, G.; Yang, Y. Planar Heterojunction Perovskite Solar Cells via Vapor-Assisted Solution Process. J. Am. Chem. Soc. 2014, 136, 622–625.

7. Jeon, N.J.; Noh, J.H.; Kim, Y.C.; Yang, W.S.; Ryu, S.; Seok, S.I. Solvent engineering for high-performance inorganic-organic hybrid perovskite solar cells. Nat. Mater. 2014, 13, 897–903.

8. Jeon, N.J.; Noh, J.H.; Yang, W.S.; Kim, Y.C.; Ryu, S.; Seo, J.; Seok, S.I. Compositional engineering of perovskite materials for high-performance solar cells. Nature 2015, 517, 476–480.

9. Seo, J.; Park, S.; Kim, Y.C.; Jeon, N.J.; Noh, J.H.; Yoon, S.C.; Seok, S.I. Benefits of very thin PCBM and LiF layers for solution-processed p-i-n perovskite solar cells. Energy Environ. Sci. 2014, 7, 2642–2646.

10. Wen, X.; Sheng, R.; Ho-Baillie, A.W.Y.; Benda, A.; Woo, S.; Ma, Q.; Huang, S.; Green, M.A. Morphology and Carrier Extraction Study of Organic-Inorganic Metal Halide Perovskite by One- and Two-Photon Fluorescence Microscopy. J. Phys. Chem. Lett. 2014, 5, 3849–3853.

11. Wang, K.-C.; Jeng, J.-Y.; Shen, P.-S.; Chang, Y.-C.; Diau, E.W.-G.; Tsai, C.-H.; Chao, T.-Y.; Hsu, H.-C.; Lin, P.-Y.; Chen, P.; et al. P-type

Mesoscopic Nickel Oxide/Organometallic Perovskite Heterojunction Solar Cells. Sci. Rep. 2014, 4.

12. Unger, E.L.; Hoke, E.T.; Bailie, C.D.; Nguyen, W.H.; Bowring, A.R.; Heumuller, T.; Christoforo, M.G.; McGehee, M.D. Hysteresis and transient behavior in current-voltage measurements of hybrid-perovskite absorber solar cells. Energy Environ. Sci. 2014, 7, 3690–3698.

13. Tress, W.; Marinova, N.; Moehl, T.; Zakeeruddin, S.M.; Nazeeruddin, M.K.; Grätzel, M. Understanding the rate-dependent J-V hysteresis, slow time component, and aging in $CH_3NH_3PbI_3$ perovskite solar cells: the role of a compensated electric field. Energy Environ. Sci. 2015, 8, 995–1004.

14. Bryant, D.; Wheeler, S.; O'Regan, B.C.; Watson, T.; Barnes, P.R.F.; Worsley, D.A.; Durrant, J. Observable Hysteresis at Low Temperature in "Hysteresis Free" Organic–Inorganic Lead Halide Perovskite Solar Cells. J. Phys. Chem. Lett.2015, 6, 3190–3194.

15. Barnes, P.R.F.; Miettunen, K.; Li, X.; Anderson, A.Y.; Bessho, T.; Grätzel, M.; O'Regan, B.C. Interpretation of optoelectronic transient and charge extraction measurements in dye-sensitized solar cells. Adv. Mater. 2013, 25, 1881–1922.

16. Snaith, H.J.; Abate, A.; Ball, J.M.; Eperon, G.E.; Leijtens, T.; Noel, N.K.; Stranks, S.D.; Wang, J.T.-W.; Wojciechowski, K.; Zhang, W. Anomalous hysteresis in perovskite solar cells. J. Phys. Chem. Lett. 2014, 5, 1511–1515.

17. Dualeh, A.; Moehl, T.; Tétreault, N.; Teuscher, J.; Gao, P.; Nazeeruddin, M.K.; Grätzel, M. Impedance spectroscopic analysis of lead iodide perovskite-sensitized solid-state solar cells. ACS Nano 2014, 8, 362–373.

18. Gottesman, R.; Haltzi, E.; Gouda, L.; Tirosh, S.; Bouhadana, Y.; Zaban, A.; Mosconi, E.; de Angelis, F. Extremely Slow Photoconductivity Response of $CH_3NH_3PbI_3$ Perovskites Suggesting Structural Changes under Working Conditions.J. Phys. Chem. Lett. 2014, 5, 2662–2669.

19. Kim, H.S.; Park, N.-G. Parameters Affecting I-V Hysteresis of $CH_3NH_3PbI_3$ Perovskite Solar Cells: Effects of Perovskite Crystal Size and Mesoporous TiO_2 Layer. J. Phys. Chem. Lett. 2014, 5, 2927–2934.

20. Koide, N.; Han, L. Measuring methods of cell performance of dye-sensitized solar cells. Rev. Sci. Instrum. 2004, 75.

21. Christians, J.A.; Manser, J.S.; Kamat, P.V. Best Practices in Perovskite Solar Cell Efficiency Measurements. Avoiding the Error of Making Bad Cells Look Good. J. Phys. Chem. Lett. 2015, 6, 852–857.

22. Zhang, Y.; Liu, M.; Eperon, G.E.; Leijtens, T.; McMeekin, D.P.; Saliba, M.; Zhang, W.; de Bastiani, M.; Petrozza, A.; Herz, L.; et al. Charge

selective contacts, mobile ions and anomalous hysteresis in organic-inorganic perovskite solar cells. Mater. Horiz. 2015, 2, 315–322.

23. Wojciechowski, K.; Stranks, S.D.; Abate, A.; Sadoughi, G.; Sadhanala, A.; Kopidakis, N.; Rumbles, G.; Li, C.-Z.; Friend, R.H.; Jen, A.K.-Y.; et al. Heterojunction modification for highly efficient organic-inorganic perovskite solar cells. ACS Nano 2014, 8, 12701–12709.

24. Shao, Y.; Xiao, Z.; Bi, C.; Yuan, Y.; Huang, J. Origin and elimination of photocurrent hysteresis by fullerene passivation in $CH_3NH_3PbI_3$ planar heterojunction solar cells. Nat. Commun. 2014, 5.

25. Xu, J.; Buin, A.; Ip, A.H.; Li, W.; Voznyy, O.; Comin, R.; Yuan, M.; Jeon, S.; Ning, Z.; McDowell, J.J.; et al. Perovskite-fullerene hybrid materials suppress hysteresis in planar diodes. Nat. Commun. 2015, 6.

26. Xue, Q.; Hu, Z.; Liu, J.; Lin, J.; Sun, C.; Chen, Z.; Duan, C.; Wang, J.; Liao, C.; Lau, L.W.M.; et al. Highly efficient fullerene/perovskite planar heterojunction solar cells via cathode modification with an amino-functionalized polymer interlayer. Mater. Chem. A 2014, 2, 19598–19603.

27. Jeng, J.-Y.; Chiang, Y.-F.; Lee, M.-H.; Peng, S.-R.; Guo, T.-F.; Chen, P.; Wen, T.-C. $CH_3NH_3PbI_3$ perovskite/fullerene planar-heterojunction hybrid solar cells. Adv. Mater. 2013, 25, 3727–3732.

28. Im, S.H.; Heo, J.-H.; Han, H.J.; Kim, D.; Ahn, T. Hysteresis-less inverted $CH_3NH_3PbI_3$ planar perovskite hybrid solar cells with 18.1% power conversion efficiency. Energy Environ. Sci. 2015, 8, 1602–1608. [Google Scholar]

29. Wu, C.-G.; Chiang, C.-H.; Tseng, Z.-L.; Nazeeruddin, M.K.; Hagfeldt, A.; Grätzel, M. High efficiency stable inverted perovskite solar cells without current hysteresis. Energy Environ. Sci. 2015, 8, 2725–2733.

30. O'Regan, B.C.; Barnes, P.R.F.; Li, X.; Law, C.; Palomares, E.; Marin-Beloqui, J.M. Optoelectronic Studies of Methylammonium Lead Iodide Perovskite Solar Cells with Mesoporous TiO_2: Separation of Electronic and Chemical Charge Storage, Understanding Two Recombination Lifetimes, and the Evolution of Band Offsets during J-V Hysteresis. J. Am. Chem. Soc. 2015, 137, 5087–5099.

31. Shkrob, I.A.; Marin, T.W. Charge Trapping in Photovoltaically Active Perovskites and Related Halogenoplumbate Compounds. J. Phys. Chem. Lett. 2014, 5, 1066–1071.

32. Stranks, S.D.; Burlakov, V.M.; Leijtens, T.; Ball, J.M.; Goriely, A.; Snaith, H.J. Recombination Kinetics in Organic-Inorganic Perovskites: Excitons, Free Charge, and Subgap States. Phys. Rev. Appl. 2014, 2.

33. Frost, J.M.; Butler, K.T.; Walsh, A. Molecular ferroelectric contributions to anomalous hysteresis in hybrid perovskite solar cell. APL Mater. 2014, 2.

34. Chen, H.-W.; Sakai, N.; Ikegami, M.; Miyasaka, T. Emergence of Hysteresis and Transient Ferroelectric Response in Organo-Lead Halide Perovskite Solar Cells. J. Phys. Chem. Lett. 2015, 6, 164–169.

35. Kutes, Y.; Ye, L.; Zhou, Y.; Pang, S.; Huey, B.D.; Padture, N.P. Direct Observation of Ferroelectric Domains in Solution-Processed $CH_3NH_3PbI_3$ Perovskite Thin Films. J. Phys. Chem. Lett. 2014, 5, 3335–3339.

36. Stroppa, A.; di Sante, D.; Barone, P.; Bokdam, M.; Kresse, G.; Franchini, C.; Whangbo, M.-H.; Picozzi, S. Tunable ferroelectric polarization and its interplay with spin-orbit coupling in tin iodide perovskites. Nat. Commun. 2014, 5.

37. Liu, S.; Zheng, F.; Koocher, N.Z.; Takenaka, H.; Wang, F.; Rappe, A.M. Ferroelectric Domain Wall Induced Band Gap Reduction and Charge Separation in Organometal Halide Perovskites. J. Phys. Chem. Lett. 2015, 6, 693–699.

38. Betoluzzi, L.; Sánchez, R.S.; Liu, L.; Han, H.; Mora-Seró, I.; Bisquert, J.; Park, N.-G.; Lee, J.-W.; Mas-Marzá, E. Cooperative kinetics of depolarization in $CH_3NH_3PbI_3$ perovskite solar cells. Energy Environ. Sci. 2015, 8, 910–915.

39. Frost, J.M.; Butler, K.T.; Brivio, F.; Hendon, C.H.; van Schilfgaarde, M.; Walsh, A. Atomistic origins of high-performance in hybrid halide perovskite solar cells. Nano Lett. 2014, 14, 2584–2590.

40. Wei, J.; Zhao, Y.; Li, H.; Li, G.; Pan, J.; Xu, D.; Zhao, Q.; Yu, D. Hysteresis Analysis Based on the Ferroelectric Effect in Hybrid Perovskite Solar Cells. J. Phys. Chem. Lett. 2014, 5, 3937–3945.

41. Xiao, Z.; Yuan, Y.; Shao, Y.; Wang, Q.; Dong, Q.; Bi, C.; Sharma, P.; Gruverman, A.; Huang, J. Giant switchable photovoltaic effect in organometal trihalide perovskite devices. Nat. Mater. 2014, 14, 193–198.

42. Zhao, Y.; Liang, C.; Zhang, H.M.; Li, D.; Tian, D.; Li, G.; Jing, X.; Zhang, W.; Xiao, W.; Liu, Q.; et al. Anomalously large interface charge in polarity-switchable photovoltaic devices: An indication of mobile ions in organic–inorganic halide perovskites. Energy Environ. Sci. 2015, 8, 1256–1260.

43. Yang, T.-Y.; Gregori, G.; Pellet, N.; Grätzel, M.; Maier, J. The Significance of Ion Conduction in a Hybrid Organic-Inorganic Lead-Iodide-Based Perovskite Photosensitizer. Angew. Chemie 2015, 54, 7905–7910.

44. Kim, H.-S.; Mora-Sero, I.; Gonzalez-Pedro, V.; Fabregat-Santiago, F.; Juarez-Perez, E.J.; Park, N.-G.; Bisquert, J. Mechanism of carrier accumulation in perovskite thin-absorber solar cells. Nat. Commun. 2013, 4.

45. Azpiroz, J. M.; Mosconi, E.; Bisquert, J.; De Angelis, F. Defect migration in methylammonium lead iodide and its role in perovskite solar cell operation. Energy Environ. Sci. 2015, 8, 2118–2127.

46. Eames, C.; Frost, J.M.; Barnes, P.R.F.; O'Regan, B.C.; Walsh, A.; Islam, M.S. Ionic transport in hybrid lead iodide perovskite solar cells. Nat. Commun. 2015, 6.

47. Mizusaki, J.; Arai, K.; Fueki, K. Ionic conduction of the perovskite-type halides. Solid State Ionics 1983, 11, 203–211.

48. Hoke, E.T.; Slotcavage, D.J.; Dohner, E.R.; Bowring, A.R.; Karunadasa, H.I.; McGehee, M.D. Reversible photo-induced trap formation in mixed-halide hybrid perovskites for photovoltaics. Chem. Sci. 2014, 6, 613–617.

49. Shi, D.; Adinolfi, V.; Comin, R.; Yuan, M.; Alarousu, E.; Buin, A.; Chen, Y.; Hoogland, S.; Rothenberger, A.; Katsiev, K.; et al. Low trap-state density and long carrier diffusion in organolead trihalide perovskite single crystals. Science2015, 347, 519–522.

50. Buin, A.; Pietsch, P.; Xu, J.; Voznyy, O.; Ip, A.H.; Comin, R.; Sargent, E.H. Materials processing routes to trap-free halide perovskites. Nano Lett. 2014, 14, 6281–6286.

51. Yin, W.-J.; Shi, T.; Yan, Y. Unusual defect physics in $CH_3NH_3PbI_3$ perovskite solar cell absorber. Appl. Phys. Lett. 2014,104.

52. Zhang, H.M.; Liang, C.; Zhao, Y.; Sun, M.; Liu, H.; Liang, J.; Li, D.; Zhang, F.; He, Z. Dynamic interface charge governing the current-voltage hysteresis in perovskite solar cells. Phys. Chem. Chem. Phys. 2015, 17, 9613–9618.

53. Ono, L.K.; Raga, S.R.; Wang, S.; Kato, Y.; Qi, Y. Temperature-dependent hysteresis effects in perovskite-based solar cells. J. Mater. Chem. A 2014, 3, 9074–9080.

54. Christoforo, M.G. I-V-Vs-Time-Taker (Python Scan Routine For Keithley 2400, Firmware Version V33). Available online: https://github.com/greysAcademicCode/i-v-vs-time-taker (accessed on 23 November 2015).

55. Christoforo, M.G. Hystanalysis (MATLAB Script). Available online: https://github.com/greysAcademicCode/hystAnalysis (accessed on 23 November 2015).

56. Unger, E.L.; Bowring, A.R.; Tassone, C.J.; Pool, V.; Gold-Parker, A.; Cheacharoen, R.; Stone, K.H.; Hoke, E.T.; Toney, M.F.; McGehee, M.D. Chloride in lead-chloride derived organo-metal halides for perovskite-absorber solar cells.Chem. Mater. 2014, 26, 7158–7165.

57. Nguyen, W.H.; Bailie, C.D.; Unger, E.L.; McGehee, M.D. Enhancing the hole-conductivity of spiro-OMeTAD without oxygen or lithium salts by using spiro(TFSI)$_2$ in perovskite and dye-sensitized solar cells. J. Am. Chem. Soc. 2014, 136, 10996–11001.

58. Leguy, A.M.A.; Frost, J.M.; McMahon, A.P.; Sakai, V.G.; Kochelmann, W.; Law, C.; Li, X.; Foglia, F.; Walsh, A.; O'Regan, B.C.; et al. The dynamics of methylammonium ions in hybrid organic-inorganic perovskite solar cells. Nat. Commun. 2015, 6.

59. Walsh, A. Principles of Chemical Bonding and Band Gap Engineering in Hybrid Organic-Inorganic Halide Perovskites. J. Phys. Chem. C 2015, 119, 5755–5760.

60. Leijtens, T.; Hoke, E.T.; Grancini, G.; Slotcavage, D.J.; Eperon, G.E.; Ball, J.M.; de Bastiani, M.; Bowring, A.R.; Martino, N.; Wojciechowski, K.; et al. Mapping Electric Field-Induced Switchable Poling and Structural Degradation in Hybrid Lead Halide Perovskite Thin Films. Adv. Energy Mater. 2015, 5.

Chapter 3

SMART LASER INTERFEROMETER WITH ELECTRICALLY TUNABLE LENSES FOR FLOW VELOCITY MEASUREMENTS THROUGH DISTURBING INTERFACES

Jürgen W. Czarske , Hannes Radner, Christoph Leithold and Lars Büttner

TUD Laboratory for Measurement and Testing Techniques, TUD Faculty of Electrical and Computer Engineering, TUD School of Engineering, TU Dresden, Helmholtzstraße 18, 01069 Dresden, Germany

ABSTRACT

Interferometric velocity measurements are of great importance at flow investigations. However, the laser beams can be distorted at the interfaces between optical media of different refractive indices. Temporal fluctuations of these distortions will cause a deterioration of the laser interferometer signals. We have harnessed the power of programmable photonics devices to eliminate this signal deterioration. Non-invasive flow velocity measurements through a rapidly fluctuating media interface with large strokes of about 100 microns are presented. Our work represents a paradigm shift for interferometric velocity measurement techniques from using static to dynamic optical elements.

MOTIVATION

Laser interferometers have become a well-established and indispensable tool for precision measurements in a huge variety of scientific and industrial applications, especially in fluid mechanics. Common measurands are displacement and velocity, which are proportional to the interference signal phase and frequency, respectively. It is well known that temperature, pressure and concentration variations of the ambient air change the refractive index [1]. Typical examples in the area of fluid mechanics are temperature gradients in combustions and pressure gradients at shock waves in compressible fluids. Beside such refractive index gradients also fluctuating interfaces between two media of different optical density have to be considered. Typical examples in fluid mechanics are water channel flows with an open surface, multi-phase flows, blood flows, levitated oscillating droplets and flows alongside a phase boundary. These variations cause severe distortions of the optical wavefronts, which can result to a failure or not tolerable uncertainties of the measurements.

To overcome these challenging metrological problems the employment of adaptive optics (AO) for a smart laser interferometer is proposed. AO is predominantly known from astronomy, where it has boosted the capabilities of modern earth-bound telescopes [2,3,4,5]. In general, an AO system comprises a wavefront sensor to measure the distortion of the wavefronts, an electric control circuit and an optical modulator to correct the distortions. Wavefront distortions caused by turbulent fluctuations of the earth's atmosphere are compensated resulting in a nearly diffraction-limited performance. A further prominent area of AO is ophthalmology. The correction of the eye aberrations enables to resolve single cells in the retina e.g., by using confocal microscopy [6] or optical coherence tomography (OCT) [7].

Also in technical metrology AO has been employed already. At optical roughness, measurements of technical surfaces the aberrations along the propagation of the laser beam have been corrected [8,9]. Also for flow measurements AO has been used [10,11,12]. First approaches have been demonstrated at laser Doppler velocimetry (LDV). AO was inserted to compensate for the random measurement deviation due to temperature gradients of heated air [10]. The LDV has been used to study gas flow stratifications.

Recently, modern AO systems have been employed for an interferometric LDV [11,12]. The deployed AO devices were supplied by Flexible Optical B.V. (OKO tech, The Netherlands) and consist of a deformable membrane mirror with integrated tip-tilt stage, a Hartmann-Shack wavefront sensor and a controller based on a PC. This adaptive LDV provide flow velocity measurements through fluctuating gas-liquid interfaces inside a water basin or channel with an open surface [11]. The distortions of the propagating laser beams due to the water surface waves have been corrected successfully, but to wave heights of only a few 10 microns [12]. At some tasks in fluid mechanics higher wave heights appear, resulting in larger amplitudes of optical distortions. In this paper we present a different approach for an adaptive, smart, interferometric LDV to overcome the former limitations. A novel electrically tunable lens and a commercially available galvanometer steering mirror are used to provide a large stroke for the wavefront correction.

For the first time, we investigate the hypothesis that electrically tunable lenses and steering mirrors enable the correction of large optical distortions in interferometry. The presented smart LDV interferometer will enable accurate flow velocity measurements through disturbing media interfaces especially at fluctuating gas-liquid interfaces. It faces large optical wavefront distortions caused by high surface waves of liquid flows, e.g., at film cooling processes, seawater salt removal, fractional distillation or processes to cool reactors.

VELOCITY MEASUREMENTS THROUGH A FLUCTUATING GAS-LIQUID INTERFACE

Preconsiderations

The flow velocity is gathered by an interferometric LDV technique, which allows high spatial and temporal resolutions together with a low velocity measurement uncertainty. The basic principle is to intersect two coherent laser beams under a small angle, forming a Mach-Zehnder interferometer. In the volume of intersection, a system of almost parallel interference fringes develops. If a particle carried with the flow to be investigated passes through the intersection volume of the laser beams, it scatters light that is modulated in amplitude with the Doppler frequency f. The velocity component v_x perpendicular to the interference fringe system orientation is resulting to

$$v_x = f \cdot d$$
(1)

with the measured Doppler frequency f [13]. The a-priori calibrated interference fringe spacing d is calculated by

$$d = \frac{\lambda}{2\sin\theta}(1 + \frac{z\cos^2\theta(z\cos^2\theta - z_W)}{z_R^2\cos^2\theta - z_W(z\cos^2\theta - z_W)})$$
(2)

with λ as the laser wavelength, θ as the intersection half angle of the interfering laser beams, z as the axis along the bisecting line of the two laser beams, z_W as the axial shift of the beam waist position relative to the center of the measurement volume and z_R as the Rayleigh length of the Gaussian laser beams, which is defined by $z_R = \pi w_0^2/\lambda$, where w_0 is the radius of laser beam waists [13].

Considering the propagation of the laser beams through an open water surface to measure the flow velocity inside the water basin, the measurement properties of an LDV are changed by a moving air-water interface. In order to evaluate the contributions of different effects, the height function $h(x,t)$ of the air-water interface is developed into a Taylor series. This procedure is in analogy to the representation with Zernike polynomials. For simplification, only the one-dimensional expression is considered:

$$h(x,t) = \underbrace{h(x_0,t)}_{Stroke} + \underbrace{(x-x_0)\frac{\partial h(x,t)}{\partial x}\bigg|_{x=x_0}}_{Tilt} + \underbrace{\frac{1}{2}(x-x_0)^2\frac{\partial^2 h(x,t)}{\partial x^2}\bigg|_{x=x_0}}_{Curvature} + ...$$

(3)

The optical distortions caused by the fluctuating air-liquid interface can deteriorate the measurement properties as follows:

- 0th order: Height or stroke of the air-water interface. For a lift of the interface a parallel shift of the beam occurs whereas the beam direction remains constant. The consequence is a shift of the position of the measurement volume,*i.e.*, a dislocation of the measurement position. Furthermore a signal frequency jitter is introduced, if a stochastic fluctuation of the air-water interface occurs.

- 1st order: Tilt of the air-water interface. Due to refraction, a tilt of the interface will change the propagation direction of the laser beam. Going to the two-dimensional consideration, it has to be distinguished between a tilt in the plane spanned by the two partial beams δ_x and a tilt in the direction normal to this plane δ_y.

- Tilt δ_x in x-direction (tip): On one side a displacement of the measurement position results and on the other side there is a change in the intersection half angle θ of the interfering laser beams. Due to a change of the Rayleigh length z_R or the beam waist position z_w a variation of interference fringe spacing d results, see Equation (2). The standard deviation σ_d represents these variations. Using Equation (1) and the propagation law of statistical independent measurement, uncertainties of the fringe spacing σ_d and the signal frequency σ_f the relative velocity measurement uncertainty yields to [13]

$$\frac{\sigma_{v_x}}{v_x} = \sqrt{\left(\frac{\sigma_f}{f}\right)^2 + \left(\frac{\sigma_d}{d}\right)^2}$$

(4)

- Tilt δ_y in y-direction: A beam deflection in the y-direction normal to the plane spanned by the two Gaussian beams will result in skew rays. The reduced overlap of the partial laser beams results in lower interference visibility and signal-to-noise ratio (SNR). Only measurement signals with a sufficient SNR are considered for further evaluation. The corresponding validation rate is given by the ratio between the evaluable and all signals detected. It represents a crucial figure-of-merit of the LDV system, since on the one hand the maximum frequency bandwidth of the velocity fluctuations and on the other the effective measurement time is determined. In the worst case of negligible laser beam overlapping no measurements can be performed at all, represented by a validation rate of zero.

- 2nd order: Parabolic curvature of the interface. Due to refraction, a curvature of the interface induces a lens effect on the beam propagation. It changes the radius of the beam waist w_0 and the position z_W of the beam waist. As a consequence, the fringe spacing d is changed according to Equation (2), which in general enhances the velocity measurement uncertainty, see Equation (4).

- 3nd order and higher orders: Distortions of the surface with high spatial frequency. The wavefront of the laser beams will be locally distorted, leading to inhomogeneities in the interference fringes. The fringe spacing can vary in all three directions. In consequence, the measurement uncertainty will increase, see Equation (4).

It should be noted that the different distortion orders strongly depend on the incidence point of the beam with respect to the surface wave, *i.e.*, on the phase of the wave. If the beam passes through a maximum of the water wave, a convergent lens will mainly affect the beam, if it passes through a minimum, a divergent lens effect appears. If the beam propagates through the zero-crossing of the wave, mainly a beam deflection and a phase jitter will occur.

A previous experimental analysis revealed that low-order Taylor series orders, respectively Zernike polynomials (piston/stroke, tip/tilt and defocus) dominate the wavefront distortions [12]. Distortions of higher orders, *i.e.*, with higher spatial frequency can be neglected at the considered flow measurement task. This can be easily understood since at the water surface the diameter of the laser beams is around some millimeters and therefore significantly smaller than the typical wavelength of the observed capillary waves of 50 mm. This wavelength results by considering the highest amplitude (mode) of the measured Fourier spectrum of 6 Hz and the estimated phase velocity of the surface wave of 0.3 m/s.

These preconditions define the spatial and temporal parameters of the AO system to be implemented. In general, several optical modulators are available for the correction of wavefront distortions, e.g., deformable membrane mirrors, micro-electro-mechanic-system (MEMS) mirror arrays and liquid-crystal spatial light modulators. In our former research work [11,12] we have employed a deformable membrane mirror exhibiting 17 electrostatic actuator elements and built-in piezoelectric tip-tilt unit. An obstacle of the deformable membrane mirror was its restricted defocus range due to the limited curvature of the membrane. Considering an optical imaging magnification from the mirror to the water surface of one, only defocuses up to about 0.3 diopters can be corrected. It is not enough at large heights of the surface waves of the investigated experiment. In the next section we will present a new optical measurement system able to handle the correction of large distortions.

Smart Laser Interferometer with Electrically Tunable Lenses

In order to correct large wavefront distortions exhibiting a strongly limited spatial bandwidth only the first and second orders of the Taylor series are considered, see Equation (3). The approach is to employ a combination of electrically tunable lenses, used to correct the 2nd order distortion and two-axis galvanometric mirrors, used to correct the 1st order distortions. Electrically tunable lenses have been suggested and employed for several measurement tasks already [14,15,16,17]. In this paper a novel adaptive silicone membrane lens, provided by Prof. U. Wallrabe (University of Freiburg, IMTEK, Laboratory for Microactuators, Germany) is inserted. The used principle is based on optofludics [14]. A piezo-electric actuator is applied for the variation of the pressure of a fluid. It results in a change of the shape of a transparent silicon membrane. In consequence, the focal length of the lens can be steered by an electrical voltage of the piezo-electric actuator [14,17]. The employed electrically tunable membrane lens offers several advantages:

- A high defocus range of up to ±40 diopters can be covered.
- A 3 dB temporal bandwidth of over 60 Hz. It should be high enough to correct the distortions of the capillary waves at the current experiment.
- Due to its outer size of 20 mm diameter the tunable lens is ideal for being integrated into compact sensors.

To correct for 1st order aberrations, i.e., tip/tilt, two biaxial two-dimensional, electromagnetically driven steering mirror (Model OIM101, Optics in Motion LLC, 4223 Rutgers Ave, Long Beach, CA 90808, USA) were used. It enables scanning over an angular range of ±1.5° (+/−26 mrad) with <2 μrad resolution and a 3 dB bandwidth of >550 Hz. Two Hartmann-Shack camera sensors (type Basler—piA640-210gm, supplied by Flexible Optical B.V., Polakweg 10–11, 2288 GG Rijswijk ZH, The Netherlands) were employed, one for each transmitted laser beam, to measure the wavefront distortions caused by the capillary waves on the free water surface. In Figure 1 the setup of the smart LDV interferometer is shown.

Based on the wavefront data of the Hartmann-Shack camera, the following actuating variables have to be computed to guarantee a complete control of the two partial beams:

- The focal lengths of the tunable lens for the first and second partial beams

- The deflections of the mirror in x-direction for the first and second partial beams
- The deflections of the mirror in y-direction for the first and second partial beams

Before the adjustment control can operate a calibration has to be performed. Therefore electric signals are applied to the control system in the absence of the distortion, *i.e.*, a silent air-water interface of the basin. The resulting changes of the wavefront are evaluated and the relationship between signals and wavefront deformation are stored as a transfer matrix. During the operation of the control loop the inverse procedure is used. The deviation of the measured wavefront from the default wavefront is calculated and multiplied with the inverse matrix to determine the electric signals which compensate the wavefront distortion. An integral controller is used as the control type, resulting in a vanishing control deviation. The control loop is realized as a twin system whose identical parts control both laser beams simultaneously. The number of performed control cycles per time unit depends on different parameters of the control loop. This control rate is mainly influenced by the number of evaluated spots of the Hartmann-Shack wavefront sensor. In the case of four evaluated spots a control rate of 1 kHz is achieved. In order to gain a sufficient sensitivity regarding the wavefront distortion, usually 20 spots are evaluated. Here the control loop operates at 540 Hz. Mainly due to the settling time of the PC-based controller, hence the latency time distortions with frequencies up to 42 Hz could be suppressed reliably. This is fast enough to cover the main components of the spectrum of the occurring capillary waves [12].

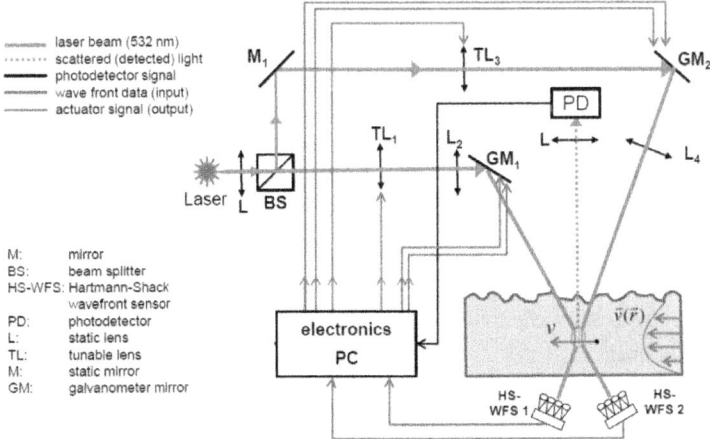

Figure 1: Smart laser Doppler velocimeter (LDV) interferometer with two implement-ed electrically tunable lenses (TL) and two two-axis galvanometer mirrors (GM). The

smart LDV compensates the wavefront distortions caused by the fluctuating air-water interface. A twin control loop was realized to control both partial laser beams simultaneously. In the overlapping area of the two laser beams a fringe system is generated. The scattered light from tracer particles moving through this fringe system is measured by the photodetector (PD). In result the flow velocity v is determined.

Performance of the Distortion Correction

To characterize the quality of the distortion correction of the AO system, different figures of merit can be considered. A primary quantity is the interference visibility, $i.e.$, the modulation degree of the burst LDV signals. Moreover, the validation rate, $i.e.$, the number of valid high SNR measurement signals with respect to the number of the total acquired signals, can be used as a figure of merit as well. Although it depends on the interference visibility and also on the SNR, this quantity is more appropriate for the flow velocity measurements. The validation rate corresponds to the measurement uncertainty and also the temporal resolution of the flow velocity.

The experiments were performed in a water basin. A water pump was used to excite capillary surface waves in the basin which acted as the optical distortion. The distortion amplitude can be varied by changing the driving voltage of the pump. In order to investigate the LDV measurement properties a moving pinhole was used as a scattering object. Below the basin an optical chopper rotated the pinhole at a constant speed v_0. According to Equation (1) the measured Doppler signal frequency is given by $f = v_0/d$. In conclusion, variations of the signal frequency are caused mainly by changes of the fringe spacing d and therefore by the optical distortions from the fluctuating air-water interface. This procedure allows us to study the improvements by the used AO system, see Figure 2 and Figure 3.

In Figure 2 the mean validation rate is displayed in dependence of the mean amplitude of the distortion. The AO system results in a significant improvement for the validation rate. At distortion amplitudes, $i.e.$, mean surface wave heights of 75 μm the validation rate is enhanced from 12% to 92% with the AO-based distortion correction. Also, compared to the former AO technique [11,12] a significant improvement has been achieved. The hypothesis to correct large distortion strokes by electrically tunable lenses and galvanometer mirrors has been confirmed.

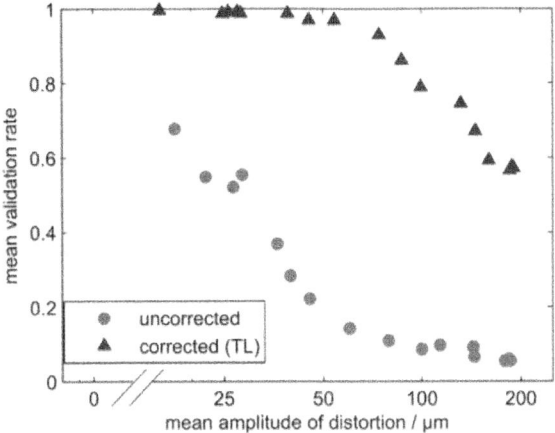

Figure 2: Mean validation rate with and without distortion correction by the tunable lens (TL) in dependence of the mean amplitude of the distortion, *i.e.*, height of the surface water wave.

As a benefit, the measurement time can be reduced or more valid signals can be gathered in a predefined measurement time. The higher data rate also increases the temporal resolution, thus making it possible to evaluate spectrally broad turbulence, which is crucial for the investigation of complex flows [18,19,20].

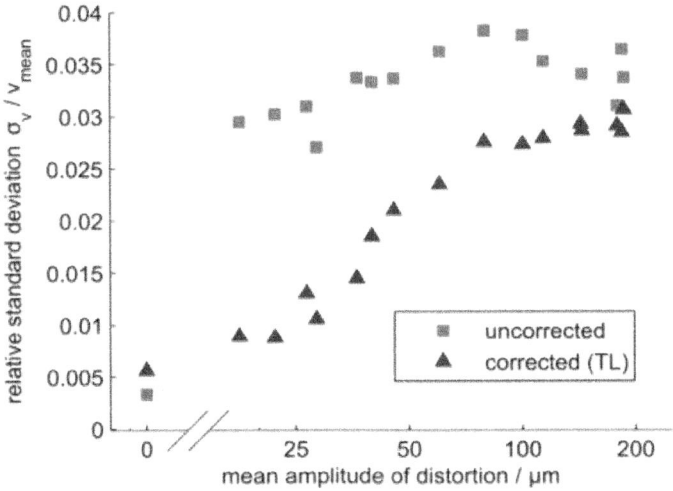

Figure 3: Relative standard deviation of the velocity measurement with and without distortion correction by the tunable lens (TL) as a function of the mean amplitude of the distortion.

Figure 3 shows the mean velocity measurement uncertainty in dependence of the mean distortion amplitude. The reason for the increased velocity standard deviation is that the spacing of the fringe system changes with the crossing angle but also with the position of the beam waist of the Gaussian beams. These effects are compensated with the tilting mirrors and the adaptive lenses, respectively. It can be seen that the velocity standard deviation is reduced significantly below mean surface wave amplitudes of about 100 μm when applying the AO correction.

Flow Velocity Measurements

A demonstration experiment was performed to proof the functionality of the wavefront correction for a real world application. A nozzle of 10 mm diameter was submerged in the water basin to generate on the one hand the flow to be measured and on the other hand capillary waves on the water surface as the optical distortion. The technical relevance is given by film flows. Since film flows exhibit a high surface-to-volume ratio, they are characterized by a high heat and mass transport. Consequently, film flows are widespread in industrial applications especially for chemical process engineering, convection processes, cleaning, evaporation, desalination of seawater, condensation and distillation.

The measurement volume of the smart interferometric LDV was located 10 mm in front of the nozzle and the flow profile was measured by mechanically traversing the nozzle. Both laser beams pass through the fluctuating water surface and are corrected. Figure 4a shows the mean velocity profile of the nozzle. The measurement uncertainty bars represent the velocity standard deviation which is mainly dominated by the turbulence of the flow. In Figure 4b,c, the mean interference contrast and the mean validation rate are displayed, respectively. It is obvious that the interference contrast is significantly improved by the wavefront correction, which consequently leads to a higher number of valid signals, *i.e.*, a higher validation rate. For a certain measurement time more valid signals can be collected which reduces the confidence interval. Both the reduced uncertainty, see Figure 3, and the higher number of valid signals, contribute to an improvement of the confidence interval as shown in Figure 4d.

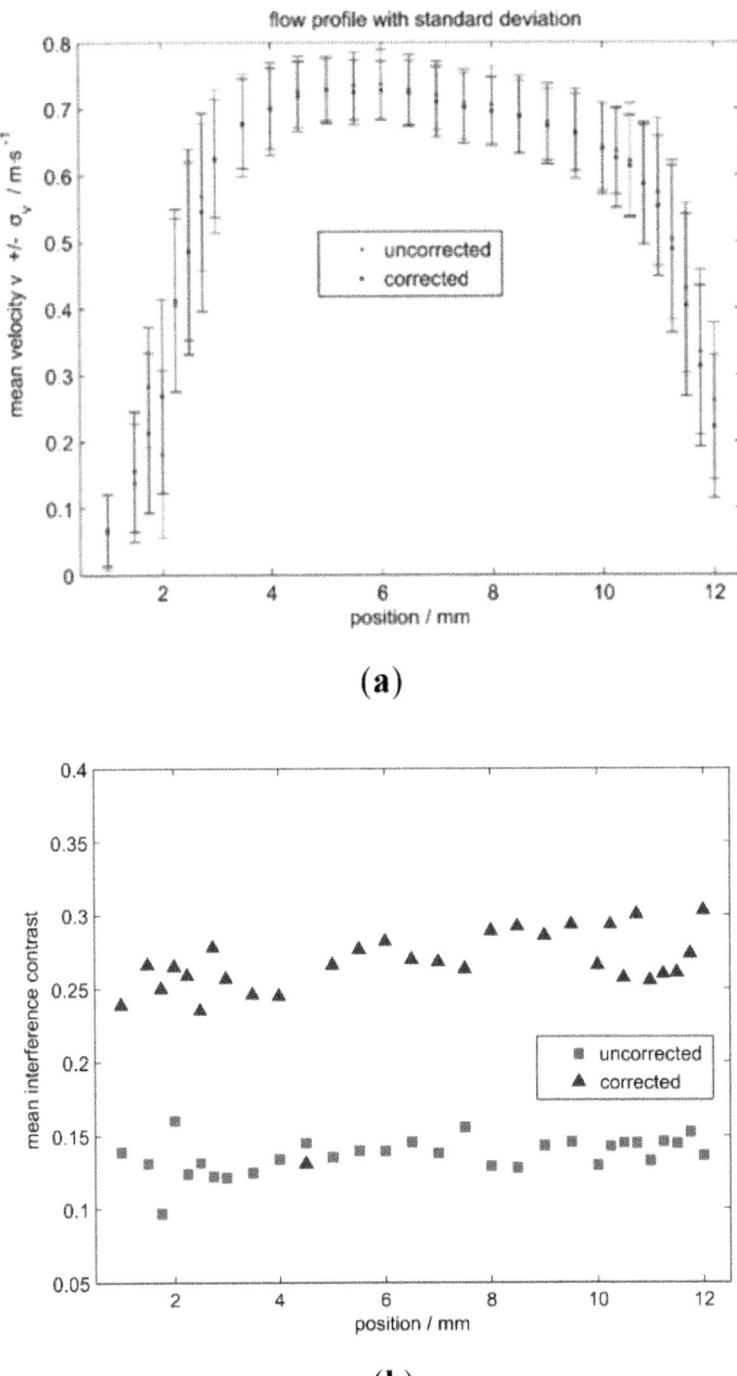

(a)

(b)

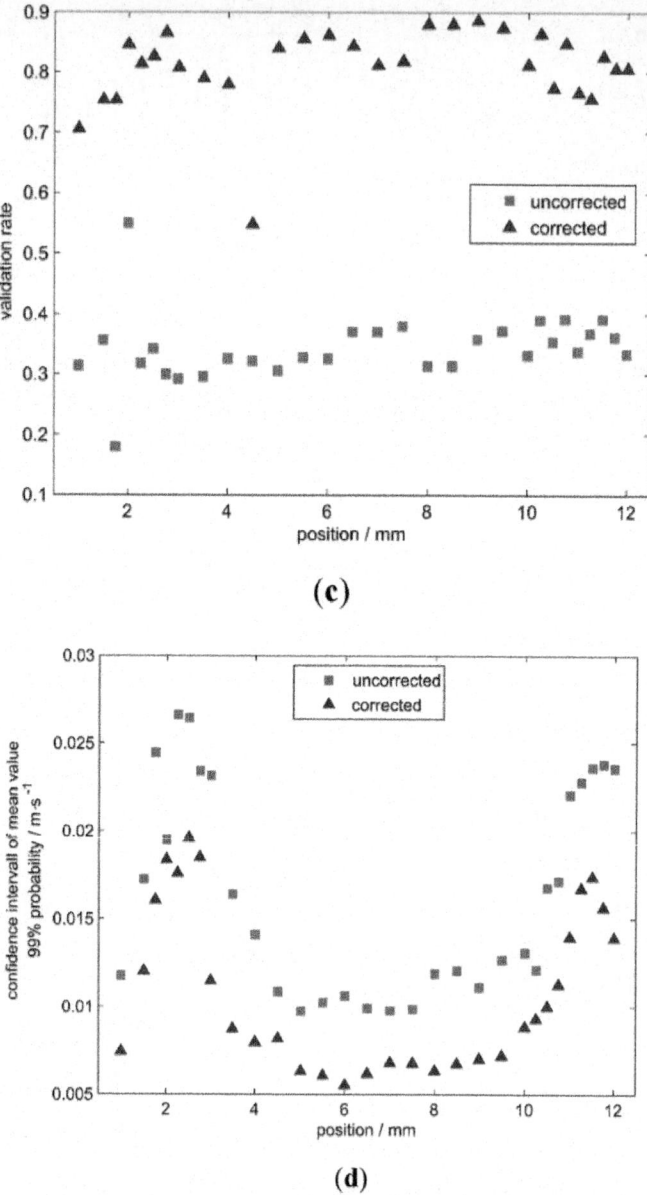

Figure 4: Measurement of the flow profile of a submerged water nozzle by the smart LDV interferometer. The experiments have been conducted for a distortion amplitude, *i.e.*, mean surface wave height of 50 μm. Both laser beams pass the free water surface, see Figure 1. They are controlled both by an adaptive optics (AO) system using electrically tunable lenses and galvanometer

mirrors. (a) Profile of the mean velocity with the velocity standard deviation as uncertainty bars. The velocity standard deviation is dominated mainly by the flow turbulence; (b) Mean interference contrast; (c) mean validation rate and (d) confidence interval, all shown as a function of the position alongside the nozzle cross-section.

CONCLUSIONS

Refractive index variations, caused by fluctuating refractive index fields or interfaces, can deteriorate the properties of velocity measurements. A lower rate of valid high SNR signals and an increased velocity uncertainty can be the consequences. To overcome these limitations, the principle of wavefront correction by means of AO was applied. AO is an emerging technology with constantly improving opportunities. Standardized fabrication processes, improved techniques and new methods significantly lower barriers of their employment. In this contribution, a smart Mach-Zehnder interferometer for flow velocity measurements was presented, that was equipped with a twin AO system to corrected wavefront distortions caused by a fluctuating air-water interface. Using the wavefront correction, the interference contrast and the number of valid high SNR signals was improved significantly. Measurements in the presence of optical distortions can be performed in a shorter time and with a higher statistical reliability. AO offers new perspectives for metrology at applications that have hardly been accessible by laser measurement techniques so far. The presented velocity measurements through disturbing fluid interfaces are important at several fields of fluid mechanics. At convection research, an improved understanding of heat transfer by turbulence can be achieved, which can enhance the energy efficiency of film cooling devices. Furthermore, in biomedical research, optical blood flow measurement can improve the investigation of diseases such as aneurysm and arteriosclerosis.

ACKNOWLEDGMENTS

The authors acknowledge financial support from the German Research Foundation (DFG, grant Cz 55/30) by a Reinhart Koselleck project. We thank M. Stürmer and U. Wallrabe (Department of Microsystems Engineering—IMTEK, Laboratory for Microactuators, University of Freiburg, Germany) for supporting us with the electrically tunable membrane lenses.

AUTHOR CONTRIBUTIONS

All authors contributed to the writing of the manuscript. H.R. performed the experimental measurements. C.L. performed numerical simulations [12], which are the basis of the design of the presented experimental setup. J.W.C. and L.B. conceived the general research direction.

REFERENCES

1.   Birch, K.P.; Downs, M.J. An Updated Edlén Equation for the Refractive Index of Air. *Metrologia* 1993, *30*, 155–162.

2.   Beckers, J.M. Adaptive Optics for Astronomy: Principles, Performance, and Applications. *Ann. Rev. Astron. Astrophys.*1993, *31*, 13–62.

3.   Hardy, J.W. *Adaptive Optics for Astronomical Telescopes*; Oxford University Press: New York, NY, USA, 1998.

4.   Glindemann, A.; Hippler, S.; Berkefeld, T.; Hackenberg, W. Adaptive Optics on Large Telescopes. *Exp. Astron.* 2000,*10*, 5–47.

5.   Tyson, R. *Principles of Adaptive Optics, Third Edition*; CRC Press: Boca Raton, FL, USA, 2010.

6.   Roorda, A.; Romero-Borja, F.W.D., III; Queener, H.; Hebert, T.; Campbell, M. Adaptive optics scanning laser ophthalmoscopy. *Opt. Express* 2002, *10*, 405–412.

7.   Hermann, B.; Fernández, E.J.; Unterhuber, A.; Sattmann, H.; Fercher, A.F.; Drexler, W.; Prieto, P.M.; Artal, P. Adaptive-optics ultrahigh-resolution optical coherence tomography. *Opt. Lett.* 2004, *29*, 2142–2144.

8.   Fuh, Y.K.; Hsu, K.C.; Fan, J.R. Rapid in-process measurement of surface roughness using adaptive optics. *Opt. Lett.*2012, *37*, 848–850.

9.   Fuh, Y.K.; Fan, J.R. Experimental investigation of a flowing fluid layer on metal surface roughness measurement and aberration correction using adaptive optics. *Opt. Rev.* 2013, *20*, 433–437.

10. Klepikov, K.E.; Kulybin, V.M.; Rinkevichius, B.S. Using the principles of adaptive optics in laser Doppler anemometry. In Proceedings of the Vsesoiuznyi Seminar po Opticheskim Metodam Issledovaniia Potokov 1st, Novosibirsk, USSR, 23–25 May 1989. (In Russian).

11. Büttner, L.; Leithold, C.; Czarske, J. Advancement of an Interferometric Flow Velocity Measurement Technique by Adaptive Optics. *Int. J. Optomech.* 2014, *8*, 1–13.

12. Büttner, L.; Leithold, C.; Czarske, J. Interferometric Velocity Measurements through a fluctuating Gas-Liquid Interface Employing Adaptive Optics. *Opt. Express* 2013, *21*, 30653–30663.

13. Czarske, J.; Büttner, L.; Razik, T.; Müller, H. Boundary layer velocity measurements by a laser Doppler profile sensor with micrometre spatial resolution. *Measurement Sci. Technol.* 2002, *13*, 1979–1989.

14. Schneider, F.; Draheim, J.; Kamberger, R.; Waibel, P.; Wallrabe, U. Optical characterization of adaptive fluidic silicone-membrane lenses. *Opt. Express* 2009, *17*, 11813–11821.

15. Khan, S.A.; Riza, N.A. Demonstration of a No-Moving-Parts Axial Scanning Confocal Microscope Using Liquid Crystal Optics. *Opt. Commun.* 2006, *265*, 461–467.

16. Riza, N.A.; Sheikh, M.; Webb-Wood, G.; Kik, P.G. Demonstration of three-dimensional optical imaging using a confocal microscope based on a liquid-crystal electronic lens. *Opt. Eng. J.* 2008, *47*, 063201:1–063201:9.

17. Koukourakis, N.; Finkeldey, M.; Stürmer, M.; Leithold, C.; Gerhardt, N.C.; Hofmann, M.R.; Wallrabe, U.; Czarske, J.W.; Fischer, A. Axial scanning in confocal microscopy employing adaptive lenses (CAL). *Opt. Express* 2014, *22*, 6025–6039.

18. Fischer, A.; Schlüßler, R.; Haufe, D.; Czarske, J. Lock-in spectroscopy employing a high-speed camera and a micro-scanner for volumetric investigations of unsteady flows. *Opt. Lett.* 2014, *39*, 5082–5085.

19. Fischer, A.; König, J.; Czarske, J.; Peterleithner, J.; Woisetschläger, J.; Leitgeb, T. Analysis of flow and density oscillations in a swirl-stabilized flame employing highly resolving optical measurement techniques. *Exp. Fluids* 2013.

20. Neumann, M.; Friedrich, C.; Kriegseis, J.; Grundmann, S.; Czarske, J. Determination of the phase-resolved body force produced by a dielectric barrier discharge plasma actuator. *J. Phys. D* 2013, *46*, 042001.

Chapter 4

MEASURING USER SIMILARITY USING ELECTRIC CIRCUIT ANALYSIS: APPLICATION TO COLLABORATIVE FILTERING

Joonhyuk Yang[1], Jinwook Kim[2], Wonjoon Kim[3], Young Hwan Kim[2]

[1]Graduate School of Culture Technology, Korea Advanced Institute of Science and Technology, Daejeon, Republic of Korea

[2]Department of Electrical Engineering, Pohang University of Science and Technology, Pohang, Republic of Korea

[3]Department of Management Science/Graduate School of Culture Technology, Korea Advanced Institute of Science and Technology, Daejeon, Republic of Korea

ABSTRACT

We propose a new technique of measuring user similarity in collaborative filtering using electric circuit analysis. Electric circuit analysis is used to measure the potential differences between nodes on an electric circuit. In this paper, by applying this method to transaction networks comprising users and items, i.e., user–item matrix, and by using the full information about the relationship structure of users in the perspective of item adoption, we overcome the limitations of one-to-one similarity calculation approach, such as the Pearson correlation, Tanimoto coefficient, and Hamming distance, in collaborative filtering. We found that electric circuit analysis can be successfully incorporated into recommender systems and has the potential to significantly enhance predictability, especially when combined with user-based collaborative filtering. We also propose four types of hybrid algorithms that combine the Pearson correlation method and electric circuit analysis. One of the algorithms exceeds the performance of the traditional collaborative filtering by 37.5% at most. This work opens new opportunities for interdisciplinary research between physics and computer science and the development of new recommendation systems.

INTRODUCTION

While various kinds of recommendation methods have been proposed, collaborative filtering (CF) is still the most widely used [1]. Typical CF techniques rely on trust or similarity relationships among users; these are based on either or both *social ties* (i.e., "I trust my friends, and the people that my friends trust") and *taste similarity* (i.e., "I trust those who agree with me") [2]. The former uses the social relationship information of users [3], which can be observed on social networking services such as Twitter [4]. The latter constructs implicit network structures for users by observing their shared behavior, such as reading the same books [5], spending time on the same Web pages [6], or purchasing the same items [7]–[10]. Some previous studies have facilitated the convergence of these two approaches through hybrid algorithms that exploit both social relationship and taste similarity [11]–[13].

In line with this, increasing attention is now being paid to network science [14], and the development of network analysis techniques is opening up new opportunities for advances in CF. This is because both social relationships and taste similarities can be easily represented as a graph or network of interconnected users. For example, [13] directly used the social network information of users to distinguish friends and strangers from their neighbors and thus enhance the CF performance. However, the utilization of social network data for CF necessarily requires additional data collection efforts. In addition, the memory-based CF with Pearson correlation (PC), Jaccard or Tanimoto coefficient (TC), or Hamming distance (HD) allows us to calculate only the direct one-to-one similarity between users instead of utilizing the information from both the direct and the indirect relationship structure. This is problematic when the number of items is far greater than the number of users and vice versa. As we observe more zeros in the user–item matrix, the similarity score between users approaches zero–this is called the sparsity problem [15].

Therefore, in this study, we propose a new technique of measuring user similarity: applying electric-circuit analysis (ECA) to the user–item matrix. By doing so, we use the full information about the relationship structure of users in the perspective of item adoption for enhanced CF that overcomes the limitations of the traditional similarity measures such as the PC, TC, or HD. ECA is used to measure the potential differences between nodes or current flow in electrical networks in physics. However, when applied to a circuit-represented consumer graph, ECA can measure the potential differences between pre-adopters and non-adopters for a product. Here, the relative potential values of non-adopters represent the similarity between users - equivalent to the PC, TC, or HD.

Based on a set of experiments, we found that the ECA algorithm exhibits better predictability than traditional user-based CF with other similarity measures. Moreover, we find that ECA and the CF methods with other measures afford different recommendation sets for items. Thus, by combining their results, we could realize a set of hybrid CF models that result in a far better recommendation performance than traditional user-based CF.

Consequently, this paper contributes to the areas of recommender systems in at least two important respects: first, instead of relying on the existing similarity calculation methods, our system employs a model that can incorporate the full relationship structure information among potential recommendation items. Second, our interdisciplinary approach between physics and computer science opens new opportunities for developing new hybrid models with possibly better performance in recommender systems. To the best of our knowledge, this is the first attempt to apply ECA to information retrieval.

THEORETICAL BACKGROUND

Network Analysis for Collaborative Filtering

CF, the most popular recommendation technique thus far, is the process of recommending relevant items to users on the basis of peer behavior [16]–[18]. In other words, CF is an assortment size-reduction process for supporting agent decisions on the basis of the choices of other agents. Because of its simplicity and powerful predictability, several firms such as Amazon.com, TiVo, Yahoo!, and Netflix have adopted the algorithm in their businesses [8],[19], [20].

Generally, CF implementations are classified into two main groups according to their method of processing data: model-based algorithms and memory-based algorithms [15]. Hybrid CF algorithms that combine memory-based and model-based algorithms constitute yet another class. Model-based algorithms aim to find behavioral patterns among users by using data-mining or machine-learning techniques. Bayesian belief net CF [21], clustering CF models [22],[23], and latent semantic CF models [24] are typical examples. Model-based algorithms usually handle the sparsity problem better than memory-based algorithms and have improved prediction performance. However, they suffer from a trade-off between scalability and predictability [15], involve expensive model-building processes, and are difficult to implement.

In contrast, memory-based algorithms are much easier to implement. The core module of these algorithms is measuring the similarity, or weight, between users. The likelihood of a user adopting or rating a product is estimated using a similarity score calculated from the weighted average of ratings from all other users. The similarity is usually calculated using the PC, TC, or HD. Because of its simplicity and ease of implementation, PC is one of the most widely adopted methods in the CF research community for quantifying the relationship between two different users [15].

One drawback of the method, however, is that the one-to-one correlation calculation cannot incorporate the relationship structure information of users. To solve this problem, there have been attempts to incorporate social network information into CF techniques [3]. Social network information enables us to observe physical interactions among users, so it helps to build better prediction models. Several studies have shown that the use of social network information can enhance the performance of CF techniques [11]–[13]. However, the requirement for additional social network data and the associated computational cost are significant limitations of such an approach.

Alternatively, [9] showed that the CF user–item matrix itself can be represented as a consumer–product bipartite graph, and explained why and when CF works using graph theory[25]: CF predictability increases with the number of n-node paths connecting consumers and products sequentially in a consumer–product bipartite graph. Although, such studies are valuable as they expand the boundaries of CF models, their important limitation is that they still largely depend on the PC based similarity measure. In other words, though the user–item matrix is represented as a network, the relationship information within the network is not effectively utilized in their approach. Here, we overcome this limitation using electric circuit theories from physics.

Electric Circuit Theory

Because electric circuit theory and analysis methods are relatively unfamiliar topics in the field of computer science, we begin with a brief introduction of ECA before discussing its connection to collaborative filtering. An electric circuit is a closed loop comprising an electrical network, which is an interconnection of electric elements (Figure 1). As a network, an electric circuit has nodes and edges. A node in an electric circuit is defined as a point where at least two electric elements are connected. There is a special node in an electric circuit called a ground node, shown as node v_6 in Figure 1. The ground node is the reference point from which other voltages are measured and is a common return path for the current.

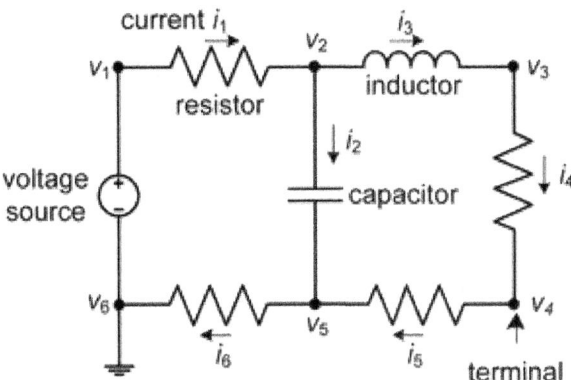

Figure 1: Typical example of an electric circuit.

doi:10.1371/journal.pone.0049126.g001

Next, electric elements are regarded as edges. There are two basic types of electric elements in a circuit: active and passive elements. First, active elements, also called sources, generate voltages or currents; that is, active elements generate energy. A voltage source generates a voltage difference between its two terminals, and a current source supplies current that is independent of the voltage across it.

Second, passive elements consume the energy generated by active elements. There are three basic passive elements: resistors, capacitors, and inductors. A resistor is an element with only resistance, i.e., zero capacitance and inductance. The resistance of an electric element, R, is the opposition to the passage of a current through the element; the inverse quantity is conductance G, i.e. the ease with which a current passes. By Ohm's law, R is defined as the ratio of voltage across an element to the current through it, and the conductance is the inverse; that is, the current-voltage (I–V) characteristic for resistance can be expressed as $V = I \times R = I \times G^{-1}$. Similarly, the capacitors and inductors also have their own I-V characteristics. For simplicity and as the first step, we utilize only the edge property of resistance in the remaining sections of this paper: this restriction could be relaxed in future studies for a more profound investigation.

Active elements generate voltages and currents that produce voltage differences and currents in the other elements. ECA is the process of finding the voltage (V) across and the current (I) through each element in the network. In other words, ECA finds the electric potential difference between two terminals of an element and the amount of current flowing through that element, which we use to measure the potential differences between pre-adopters and non-adopters for a product in our study.

ECA for Collaborative Filtering

As mentioned above, ECA identifies all the potential differences between terminals in an electrical network, given that at least one power source is connected to the network. The potential difference between two terminals is directly interpreted as the voltage value of the node. In this study, we use the voltage difference between two nodes as user similarity. That is, the voltage differences between nodes in the circuit can represent a consumer graph, thus replacing the other similarity measures.

A metaphorical example might help in understanding the principle of circuit representation and ECA. Imagine a number of marbles scattered on a flat table. The marbles all look different, and they are interconnected by elastic rubber bands; that is, they form a network. Each rubber band has its own elastic constant, which is inversely proportional to the similarity between the two marbles it connects. The more similar the two marbles are, the less elastic the rubber band is. In other words, the bands hold similar marbles together more tightly. Conversely, the band is more elastic if the marbles are less similar. In the same manner, all the marbles are also connected to the table by rubber bands. This is the same situation as that in the case of the circuit we represent in this paper. For ECA, we pick a marble in the network and lift it up to a certain height. Accordingly, the other marbles are also lifted up, following the picked marble. The height to which a following marble is lifted depends on the elasticity of all the rubber bands going from that marble to (1) the picked marble, (2) the table, and (3) all the other marbles not first picked. The more similar a following marble is to the picked one, the higher the following marble is lifted. ECA measures the exact heights of all the marbles. The following sections formally elucidate this process in detail.

MATERIALS AND METHODS

Circuit Representation

As shown in Figure 2, the user–item binary matrix can be converted into a consumer–product bipartite graph, and then projected onto a consumer graph [9]. The user–item binary matrix inFigure 2(a) shows that c_1 purchased p_1, p_2, and p_4. Similarly, p_2 is purchased by c_1 and c_2. The matrix can be drawn as a consumer–product bipartite graph, as shown in Figure 2(b). We connect a consumer to a product if the consumer purchased the product, so there are no connections between consumers or products themselves. As described in [9], we can now obtain a network of consumers with the relationships among them by projecting the edges of the bipartite graph onto the consumer side. Consumers are connected as shown in Figure 2(c)if there is at least one common purchased product among them. The double line between c_1 and c_2 indicates that they have purchased two products in common. Finally, we convert the consumer graph to an electric circuit, as shown in Figure 2(d).

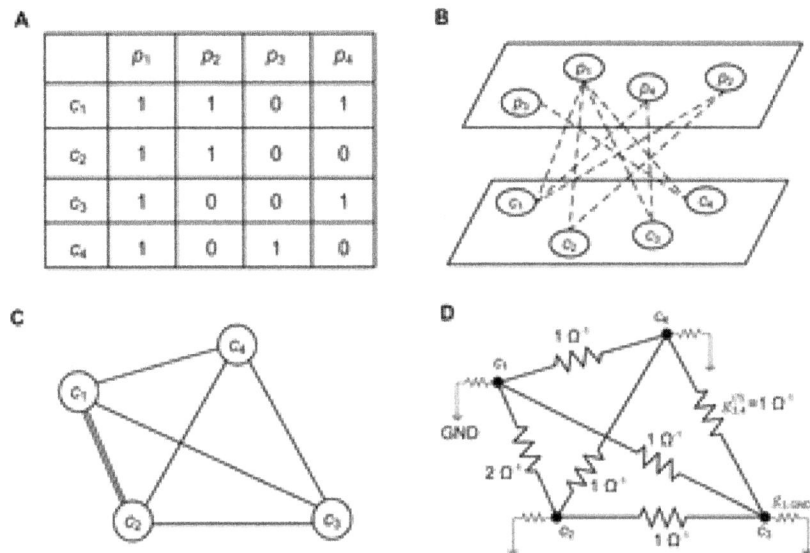

Figure 2: Circuit representation procedure.

A: user–item binary matrix; B: consumer–product bipartite graph [9]; C: projected consumer graph [9]; and D: electric circuit representation of the projected consumer graph.

doi:10.1371/journal.pone.0049126.g002

In the electric circuit, each node represents a consumer or user, the conductance $g_{i,j}$ between nodes i and j captures the taste similarity of two consumers, and the conductance $g_{i,GND}$ between node i and the ground captures a characteristic of a consumer such as the annual number of purchases. In this paper, we propose three different ways of defining the conductance, as follows:

$$g_{i,j}^{(1)} = \text{number of purchases in common,}$$

$$g_{i,j}^{(2)} = \frac{g_{i,j}^{(1)}}{\max(\deg_i, \deg_j)},$$

$$\text{and } g_{i,j}^{(3)} = \frac{g_{i,j}^{(1)}}{\text{mean}(\deg_i, \deg_j)},$$

$$(1)$$

where $\deg_i$ is the number of edges outgoing from consumer i. As $g_{i,j}^{(1)}$ is the number of common purchases between consumers i and j, it simply quantifies the degree of shared behavior of two consumers in the past. This is under the assumption that a higher frequency of shared behavior between two consumers implies a higher similarity between them. $g_{i,j}^{(2)}$ is the normalized shared behavior frequency, obtained by dividing $g_{i,j}^{(1)}$ by the maximum degrees of consumers i and j; it captures the effect whereby more actively purchasing consumers get relatively greater weights with all connected consumers. Similarly, $g_{i,j}^{(3)}$ is normalized with the average degree of consumer i or j.

A characteristic of the consumer, such as the total number of purchases, is modeled as the grounded conductance $g_{i,GND}$ of the corresponding node i. In this paper, we propose two ways of defining the grounded conductance, as follows:

$$g_{i,GND}^{(1)} = k \quad \text{and} \quad g_{i,GND}^{(2)} = k \times \deg_i^{-1}, \tag{2}$$

where k is a constant. The constant grounded conductance model assumes that all consumers are equally likely to purchase a product. The inverse degree model assumes that consumers with greater numbers of previous purchases are more likely to purchase a product. In this case, a higher grounded conductance indicates a lower voltage level for a node, i.e., a smaller likelihood for purchasing the product.

Measuring Similarity Using ECA

The first step of any recommendation algorithm is to estimate a likelihood score for each candidate product a user is likely to purchase and to sort the candidates on the basis of that score. This is usually called the top-k nearest neighbor (kNN) method. The essential role of ECA is to determine the score as a voltage level. There are two different ways of using ECA for likelihood estimation, according to which type of graph we are dealing with, i.e., a projected consumer graph or a projected product graph. We predict which consumers will purchase a specific product by applying ECA to the projected consumer graph, whereas we predict which product a consumer will purchase with the projected product graph.

Before we explain the method for applying ECA to a projected consumer graph, we need to understand two natural laws in ECA: Kirchhoff's current law (KCL) and Kirchhoff's voltage law (KVL). KCL deals with the conservation of electric charge, which implies that the sum of currents flowing into a node is equal to the sum of currents flowing out of that node. In Figure 1, at node v_2, the incoming current i_1 is equal to the sum of outgoing currents, i_2+i_3. KVL deals with the conservation of energy in an electric circuit, which implies that the directed sum of electric potential differences (voltages) around any closed circuit is zero. In Figure 1, for loop $v_1-v_2-v_5-v_6$, the sum of voltage differences between adjacent nodes, $(v_1-v_6)+(v_2-v_1)+(v_5-v_2)+(v_6-v_5)$, is equal to zero. By applying KCL and KVL to an electric circuit, we can obtain a set of equations describing all branch currents and node voltages. By solving the simultaneous equations, we can find all the node voltages and branch currents. Several methods have been developed for generating these circuit equations systematically, and modified nodal analysis [26] is commonly used.

Therefore, we can apply ECA to a projected consumer graph through the following steps. Note that the opposite case, applying ECA to a projected product graph, is just the dual, so we do not include it in this paper.

1. Represent a user–item binary matrix as a projected consumer graph.

2. Represent the projected consumer graph as an electric circuit.

3. Select a target product and connect unit voltage sources to all the consumers who have purchased the product at that time.

4. Formulate simultaneous equations for the voltage levels of other consumers and the current flow through resistors using KCL and KVL.

5. Solve the equations and obtain all the voltage levels of the consumers. The values represent the likelihood of a consumer purchasing the product: the greater is the voltage value, the higher is the likelihood.

To understand the ECA algorithm more clearly, we applied ECA to the circuit representation of the consumer graph in Figure 2 (see Table 1). With the network, let us suppose we observed that c_1 purchased a newly released p_5. Then we connect c_1 to a power source with a unit voltage, and current flows from it to all the connected consumers. The ECA measures the potential differences of all the other consumers compared with the value of 1 V for c_1. Under the configuration of $g_{i,j}^{(1)}$ conductance and constant grounded conductance, the voltage at c_2 is 0.615 V, and those at c_3 and c_4 are 0.538 V. This implies that c_2 is more likely to purchase p_5 than c_3 and c_4. This result is intuitive because c_2 is more similar to c_1 than to any other consumer in terms of the historical shared behavior. In the same manner, if we observed that c_2 purchased p_5, then the likelihoods of purchase would be 0.615 V for c_1 and 0.538 V for the other consumers under the same conditions.

Table 1: ECA results for the circuit representation of consumer graph in Figure 2(d).

	c_1	c_2	c_3	c_4
v_1	1.000 V	0.615 V	0.500 V	0.500 V
v_2	0.615 V	1.000 V	0.500 V	0.500 V
v_3	0.538 V	0.538 V	1.000 V	0.500 V
v_4	0.538 V	0.538 V	0.500 V	1.000 V

Under the assumption of edge weight $g_{i,j}^{(1)}$ and ground weight $g_{i,GND}^{(1)}$.
doi:10.1371/journal.pone.0049126.t001

doi:10.1371/journal.pone.0049126.t001

Data

We use two datasets: the Movielens dataset, available at http://www.grouplens. org/node/73, and a set of publically inaccessible book transaction logs. The 1 M MovieLens dataset contains 1,000,209 anonymous ratings of 3,952 movies by 6,040 users. The ratings are made on a five-star scale, which is inappropriate for our proposed method that is only applicable to binary data analysis at the moment. Thus, we convert all the ratings to one and null values to zero, as suggested by [27].

Second, individual-level book transaction logs were obtained from an anonymous bookstore chain. The bookstore has both nation-wide offline retailing stores and an online e-commerce Web site. The data we used cover both online and offline sales for three book categories–novels, poetry, and essays– from the first day of 2006 to the last day of 2008. This includes 9,934,309 transaction logs of 1,839,674 distinct registered members, who purchased 62,109 distinct books. Thus, we have a 1,839,674 × 62,109 user–item binary matrix. Because of computational limitations, we used a network-sampling technique, the Metropolis–Hastings random walk (MHRW) algorithm [28]. Unlike random sampling, the MHRW algorithm allows us to sample a certain number of observations without violating the degree distribution of the population. Using the algorithm, we picked 10,000 consumers who purchased more than ten books for the training set.

Experimental Design

To evaluate the predictability of the recommendation, we divided the dataset into two groups, namely, the training and test sets. For the MovieLens dataset, we used a randomly selected 80% of the books for the training set and the remaining 20% for the test set. The consumer graph of the training set consists of 6,040 nodes and 17,141,741 edges, which implies a density of 0.940. The mean degree of the graph is 5,700 with a standard deviation of 3,300. For the book transaction logs, we used the first two years, i.e., 2006 and 2007, for the training set and the last year, 2008, for the test set. The basic concept of our experiment is that we recommend to N potential consumers that they purchase a product in the test set and evaluate the recommendation results by tracking whether the consumers purchased the product. For example, Figure 3 depicts the recommendation and evaluation procedure for the book transaction logs. First, a new product is released at t_r and we wait until t_e when n consumers purchase it, so that we can determine which nodes in the circuit-represented the consumer graph with bias unit voltage. Second, we draw a circuit using the user–item binary matrix at time t_e and connect a unit-voltage power source to the n consumers who pre-purchased it. Third, by applying ECA, we obtain the voltage values for all the consumers in the circuit, and sort them in descending order. Fourth, we pick the top N consumers and make a recommendation. Fifth, we evaluate the quality of the recommendation L weeks later, i.e., at t_e+L, by determining whether the recommended consumers purchased the product during t_e or t_e+L.

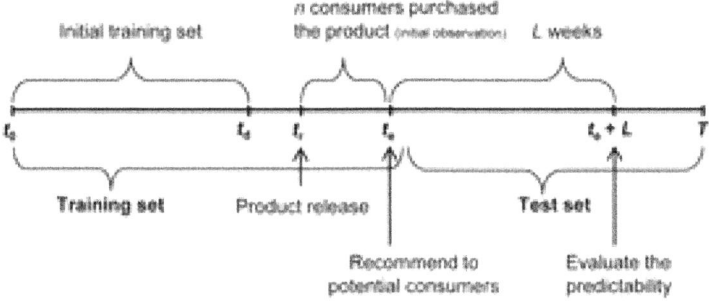

Figure 3: Experimental design.

doi:10.1371/journal.pone.0049126.g003

We used three decision-support metrics to evaluate the prediction accuracy [18], [29]: precision, recall, and F-measure. Precision is the ratio of the number of hits h to the number of recommendations m. Recall is the ratio of h to the number of actual sales S in the test set. F-measure is a combination of these two metrics.

$$P = \frac{h}{m}, \quad R = \frac{h}{S}, \text{ and } F = \frac{2 \times P \times R}{P + R}.$$

(3)

We performed the experiment for the top-ten best-selling books released between January 2008 and March 2008 for ten different sets of consumer samples. Thus, we obtained 100 precision, recall, and F-measures, and averaged them for each recommendation algorithm. A typical consumer graph of the 10,000 nodes has 14,115,000 edges, i.e. density of 0.282, and 2,400 mean degree with a standard deviation of 1,700.

RESULTS AND DISCUSSION

We begin this section by graphically showing the results of ECA on a consumer graph. Figure 4A shows a set of 500 nodes, where each node represents a consumer and five of them purchased a newly released product (in larger circles). The edge weight between two consumers is determined by the number of common products purchased by the two consumers in the past. Next, the node colors represent the potential differences between nodes. The red in the five lager nodes indicates that those nodes are connected to a power source. If the color of a node is closer to red, it has a higher voltage. Conversely, a hue closer to yellow means the node has a lower voltage. Then, Figure 4B shows the purchase status of the same consumers several weeks later. Totally, 41 nodes are in the larger circle, which means that there were 36 additional consumers who purchased the product. Note that the larger nodes are closer to red than the smaller nodes. The remainder of this section formally quantifies and discusses the recommendation quality of the proposed approach.

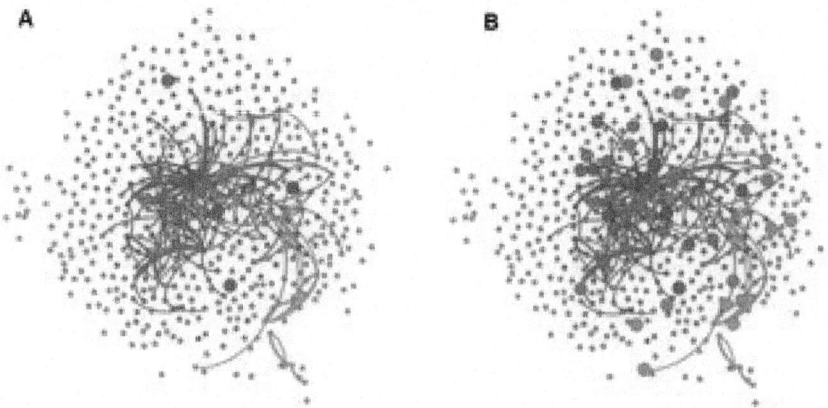

Figure 4: Result of ECA on consumer graph.

Nodes for consumers and edges for the number of shared purchases in the past between consumers. For graphical clarity, we dropped the edges with weight less than or equal to five. We used [35] for the graphic tool.

doi:10.1371/journal.pone.0049126.g004

Recommendation Quality

In order to compare the recommendation quality of ECA with that of existing methods, we used user-based CF [18], [30] with three different similarity measures: PC, TC, and HD. The benchmark model calculates the score S_i of a potential adopter i at the time point t_e as: $S_i = (1/N_p) \sum_{j \in P} w_{i,j}$, where P is a set of pre-adopters, N_p is the number of pre-adopters, and w_{ij} is the similarity between user i and j. Then we recommend the product to the top-N potential adopters by sorting S_i in descending order. We omit the well-known equations for the three measures to calculate the similarity (see, e.g., [1], [19], [31], [32]). Even though there are some recently developed variations of the basic form of the similarity measure, especially in PC, we could not apply such advancements into our benchmark models. For example, case amplification [18] does not have an effect on the quality metrics in this study because it does not change the order of similarity scores. It rather helps to improve other metrics such as the mean absolute error. The scale and transformation invariant PC [33] is another example we failed to incorporate to our benchmark model owing to its narrow applicability to only voting or rating data. However, we can still find relatively recent papers describing newly developed algorithms that could be compared with the traditional benchmark models used in this study[13], [34]. The choice of a memory-based algorithm as benchmark is attributed to this study's primary focus on the technique of similarity measure in memory-based CF, and not model-based algorithms without a similarity calculation. One might argue that relatively recent variations of the benchmark model, which generally exhibit better performance, should be used. The main purpose of this study, however, is to provide a technique for measuring similarity or proximity among users, which is the core module for a memory-based CF.

We compare the performance of the benchmark models and six ECA variations for three different inter-consumer conductance models and two different grounded models. For example, ECA1 is conducted under a configuration with $g_{i,j}^{(1)}$ conductance assumptions between consumers and constant grounded conductance $g_{i,GND}^{(1)}$.

Figure 5 shows the experimental results for the MovieLens dataset. While ECA4, with $g_{i,j}^{(2)}$ and $g_{i,GND}^{(1)}$, is the highest in all the quality measures, all six ECA variations outperform the benchmark models (i.e. PC, TC, and HD). The relatively smaller average elapsed time needed to make one recommendation is also a strength of ECA. When we varied the number of recommendations to different numbers, such as 10 or 50, to check the robustness of the results, there was no significant change in the performance rank among measures.

The results for the book transaction data, shown in Figure 6, confirm what we found in the MovieLens experiment. Figure 6 shows the *F*-measures of all measures listed in an increasing number of weeks required to predict. For example, a value of five for *L* implies that we predict the actual adoption status of consumers five weeks after the similarities among users are measured. This variation in *L* is primarily because the quality metrics, precision and recall, are list-length based. Figure 6A shows that ECA4 is the best among all the six variations as we found in the previous experiment. The gaps between different types of ECA measure do not seem to vary with an increase in *L*. ECA4 is compared with the benchmark models in Figure 6B. For better visibility in the figure, we omitted the HD results, which had far smaller *F*-measure values than the other measures (.009 to.02). While the *F*-measure of ECA4 is higher than that of PC and TC, the gaps between ECA4 and the benchmark models decrease with an increase in *L*. This suggests that ECA performs particularly well when *L* is small, which can be an advantage of ECA from practical point of view. Thus, we can conclude that the ECA measures can be successfully incorporated into recommender systems.

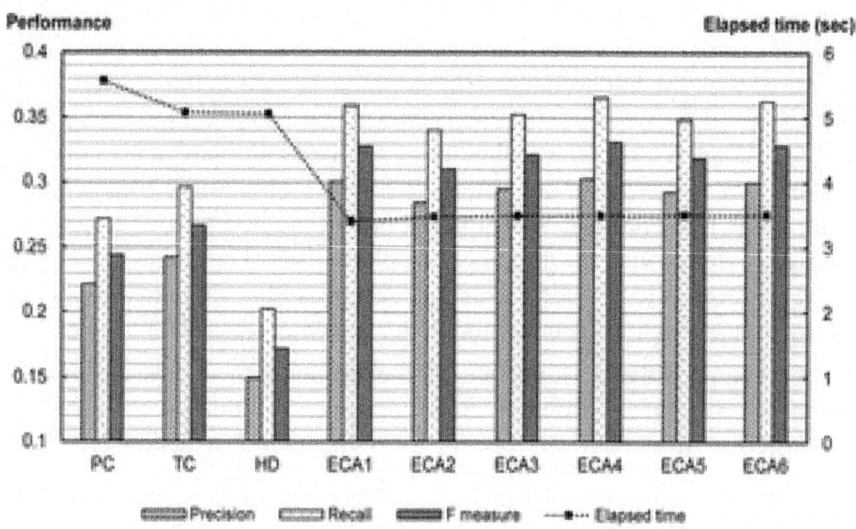

Figure 5: Experimental results for MovieLens dataset.

Number of recommendation is 100. There was no significant change in the performance rank among measures when we varied the number to 10 and 50.

doi:10.1371/journal.pone.0049126.g005

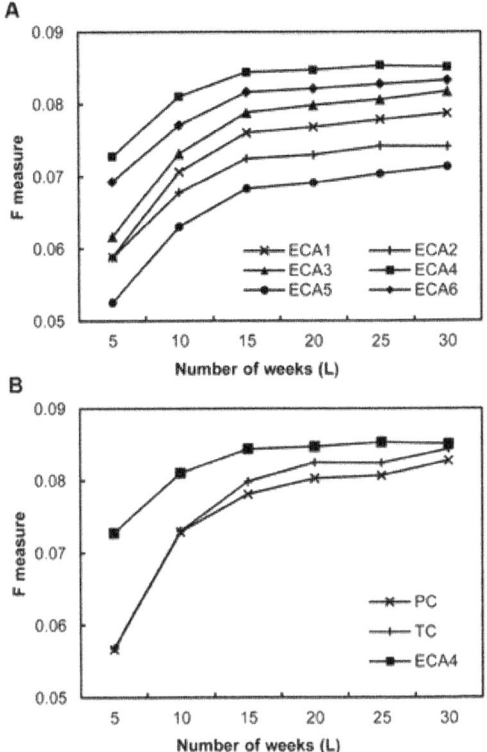

Figure 6: Experimental results for book transaction dataset.

doi:10.1371/journal.pone.0049126.g006

Graph Density Reduction

Another important quality criterion for recommendation algorithms is scalability [19]. Whereas PC computes only correlations among a node's n neighbors, ECA demands matrix inverse computation at least once. The cost for computing a matrix inverse increases with the cube of the number of nodes, and this could be a drawback of our ECA measures. As shown in Figure 5, applying ECA to time-independent dataset, such as the MovieLens dataset, does not pose a significant scalability problem. The worst case would be where the user-item matrix is updated in real-time whenever there is any additional transaction or adoption. This is problematic because it demands matrix inverse computation each time to make a recommendation. In a practical application, a recommendation system can update its user-item matrix periodically to avoid such computational costs. Moreover, the speed of ECA greatly improves as the density of the consumer graph decreases.

To check the effect of graph density reduction on the ECA prediction accuracy and computation cost, we performed an experiment in which we removed edges whose weights are less than a given threshold H and we adjusted the weights of the remaining edges such that $w'_{i,j} \sim \max (0, w_{i,j} \{H)$. The density reduction threshold was varied from 1 to 14 and we used the book transaction dataset. The other experimental settings were all the same as in the previous section, except we sampled four consumer sets instead of ten. A Linux machine with an Intel Core i7 870 quad-core processor and 8 GB of memory was used.

Figure 7 illustrates the results. The bar chart shows the average values of the F-measure and the plot represents the average computation elapsed time. The results show that the prediction accuracy of ECA decreases owing to the loss of information until the threshold reaches four, but increases again with H. Our explanation for this phenomenon is the effect of noise reduction: by removing meaningless edges, which can be viewed as noise, only meaningful edges remain and the prediction accuracy is improved. Although the F-measure did not reach the value of ECA without density reduction, the computation elapsed time is greatly reduced and this could be one of the solutions to the scalability problem of ECA. In-depth analysis of density reduction would be necessary in the future.

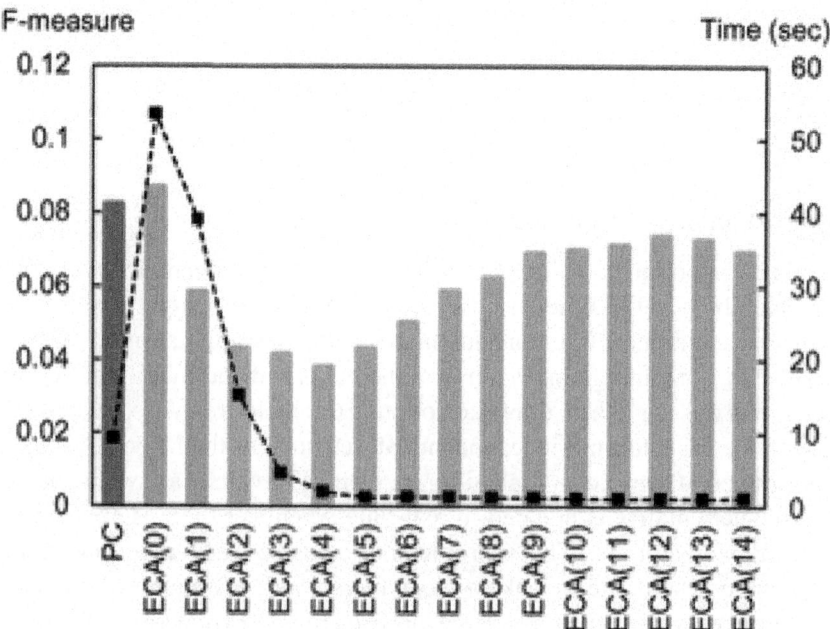

Figure 7: Result of graph density reduction.

The red bar represent the Pearson correlation method, and orange bars represent the electric circuit method. ECA(H) for graph density reduction with threshold H, and ECA(0) represents no graph reduction.

doi:10.1371/journal.pone.0049126.g007

Hybrid Approaches

One additional interesting finding of this study is that the potential consumers are given a recommendation and the hit recommendations differ among the various measures. The smaller intersection areas of the hit sets of different measures imply that there is a greater opportunity for us to improve the performance with a hybrid algorithm that recommends the union hits set. To check the feasibility of such hybrid methods, we compared the number of consumers in the intersection of PC and ECA4 recommendations and the number of hits in common. We used the book transaction dataset.

Table 2 shows the results. Only 11.7 products out of the two top-100 recommendation lists overlapped on an average, and only 3.4 recommendations were hits; that is, the PC and ECA algorithms target different sets of consumers for recommendations, and thus the results show promise for possible complementary use of the two algorithms. For the best case, we can increase the number of hit recommendations for product "1" to 43.6 (=27.9+20.2–4.5), i.e., a 56.3% increase compared with CF, if we include all the hits of PC, ECA, and the intersection of the recommendation lists.

Table 2: Feasibility check for ECA–PC hybrid algorithms

Product	A	B	C	D	E
1	27.9	20.2	10.6	4.5	43.6
2	30.2	25.6	14.7	4.9	50.9
3	17.8	14.7	12.5	3.1	29.4
4	9.9	10.7	8.1	1.1	19.5
5	10.3	11.5	8.4	1.7	20.1
6	12.6	12.2	9.1	2.8	22.0
7	8.1	7.7	11.9	1.9	13.9
8	20.3	25.5	13.6	5.2	40.6
9	2.5	2.4	8.2	0.4	4.5
10	17.3	21.1	19.4	7.9	30.5
Mean	15.7	15.2	11.7	3.4	27.5

A: number of items hit out of 100 using the Pearson correlation method; B: number of items hit out of 100 using the ECA method; C: number of the products recommended by both the Pearson correlation and ECA method; D: number of hits out of the common recommendations (C); and E: number of the best possible hits.
doi:10.1371/journal.pone.0049126.t002

doi:10.1371/journal.pone.0049126.t002

Based on the above findings, we propose four types of ECA–PC hybrid methods heuristically: (1) top-N split, (2) top-N combination, (3) top-N common, and (4) weighted average. Figure 8 depicts the first three approaches graphically. First, the top-N split approach (Figure 8A) generates two sorted consumer lists, one from PC and one from ECA, and then picks the top-N_A consumers from PC and the top-N_B consumers from ECA, where $N_P+N_E=N$. We varied the ratio in three ways: the mixing ratio of PC versus ECA for hybrid 1a (H1a) is 3:7, H1b is 5:5, and H1c is 7:3.

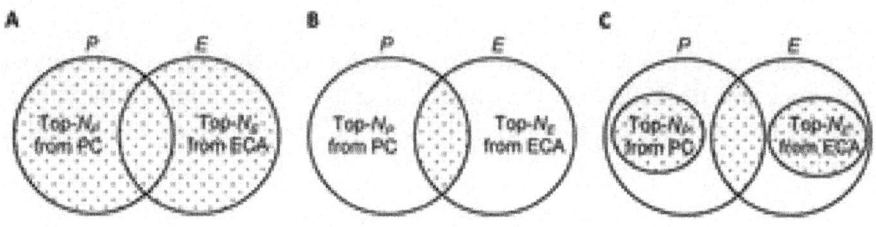

Figure 8: ECA–PC hybrid algorithms.

A: hybrid1 (H1) top-N split, $n(A \cup B)=N$, where $n(P):n(B)=a:b$ and $(a,b)=\{(3,7), (5,5), (7,3)\}$; B: hybrid2 (H2) top-$N$ combination, $n(P')+n(E')+n(P \cap E)=N$, where $P'=P-(P \cap E),E'=E-(P \cap E)$, $n(P \cap E):n(P'):n(P')=a:b:c$, and $(a,b,c)=\{(1,1,1),(2,1,1), (1,2,1),(1:1:2)\}$; and C: hybrid3 (H3) top-N common method, $n(P \cap E)=N$.

doi:10.1371/journal.pone.0049126.g008

Second, the top-N combination approach (Figure 8B) is similar to the top-N split, except that it allocates a portion to the consumers in the intersection area. The mixing ratio of intersection versus PC versus ECA is 1:1:1 for H2a, 2:1:1 for H2b, 1:2:1 for H2c, and 1:1:2 for H2d. Third, the top-N common approach (Figure 8C) picks the top-N consumers, listed in descending order, from PC and ECA.

Finally, the weighted average approach simply takes the average values for PC and ECA for a consumer with varying weights:

$$Score_i = a \cdot S(PC_i) + b \cdot S(ECA_i) \quad \forall i, \tag{4}$$

where,

$$S(PC_i) = \frac{PC_i - \text{mean}(PC_i)}{\text{stdev}(PC_i)},$$

$$S(ECA_i) = \frac{ECA_i - \text{mean}(ECA_i)}{\text{stdev}(ECA_i)}. \tag{5}$$

The PC versus ECA weight ratio (a,b) is 3:7 for H4a, 5:5 for H4b, and 7:3 for H4c.

Table 3 shows the performance evaluation results for the hybrid algorithms. Most hybrid algorithms exceed the performance of both PC and ECA4. The best performance among the hybrid algorithms is H3, top-N common. The precision of H3 is 29.1% higher and the recall is 37.5% higher than that of PC. However, we admit that it is not clear why H3 outperforms the other hybrid measures at the moment. The development of better hybrid algorithm and understanding the underlying mechanism of the different hit sets would constitute challenging and promising future work.

Table 3: Performance of ECA–PC hybrid algorithms

	Precision	Recall	F-measure
PC	0.1569	0.0562	0.0828
ECA4	0.1516	0.0592	0.0851
H1a	0.1558	0.0593	0.0799
H1b	0.1571	0.0589	0.0797
H1c	0.1559	0.0571	0.0775
H2a	0.1697	0.064	0.0866
H2b	0.1776	0.0676	0.0911
H2c	0.1642	0.0606	0.0823
H2d	0.1669	0.0639	0.0861
H3	0.2025	0.0773	0.104
H4a	0.1619	0.0585	0.0795
H4b	0.159	0.0572	0.0779
H4c	0.1579	0.0566	0.0771

doi:10.1371/journal.pone.0049126.t003

doi:10.1371/journal.pone.0049126.t003

CONCLUSIONS

The study of recommender systems is an important research area as decision makers are now faced with far more abundant information and an increasing assortment of choices. In this paper, we proposed and analyzed a new, but preliminary, algorithm that can measure the similarity between individuals in product consumption using electric circuit analysis (ECA). Specifically, we represent a consumer graph [9] as an electric circuit, and apply ECA to the circuit representation of the consumer graph. By doing so, we measure the potential differences between pre-adopters and non-adopters for a product and the value allows us to determine recommendation priorities. We also propose four distinct hybrid approaches to enhance the predictability of the algorithm by combining the recommended sets from PC and ECA.

The experimental results show that ECA can be an appropriate and performance-enhancing module for recommender systems. In particular, ECA outperforms all the benchmark models with the MovieLens dataset and one of the proposed ECA–PC hybrid algorithms surpasses the performance of traditional PC by 37.5% in recall. In addition to the development of an advanced recommendation algorithm, this paper has two important implications. First, by using ECA, we can fully employ the network information among consumers or products instead of relying on the existing one-to-one similarity calculation methods such as PC or TC. Second, by applying ECA to management information retrieval, we can open new opportunities for interdisciplinary research among different academic disciplines for resolving the problems in information science.

Several interesting research issues can be addressed in future extensions of this work. First, in addition to conductance, which we actively used in this paper, there are unexplored electric circuit elements such as capacitors and inductors. The electric characteristics of these elements are promising in terms of consumer-behavior modeling. Second, an electric circuit, as well as its analysis, is necessarily dynamic. Dynamic link prediction [10] on consumer–product graphs can be implemented using ECA. Third, one can extend the ECA measure to be applicable to rating or voting data. Fourth, as mentioned in the previous section, understanding the underlying mechanism of the different hit sets among various measures would open great opportunities to develop hybrid algorithms. Finally, although we estimated the proximity of consumers using implicit behavioral relationships in this paper, one can directly incorporate explicit social relationship data into circuit-representation processes. By doing so, a new ECA-based socio-metric index or information diffusion model could be developed.

AUTHOR CONTRIBUTIONS

Conceived and designed the experiments: JY JK. Performed the experiments: JY JK. Analyzed the data: JY JK. Contributed reagents/materials/analysis tools: WK YHK. Wrote the paper: JY JK WK YHK.

REFERENCES

1. Herlocker JL, Konstan JA, Terveen LG, Riedl JT (2004) Evaluating collaborative filtering recommender systems. ACM T INFORM SYST 22: 5–53. doi: 10.1145/963770.963772

2. Dell'Amico M, Capra L (2010) Dependable filtering: Philosophy and realizations. ACM T INFORM SYST 29: 5. doi: 10.1145/1877766.1877771

3. Kautz H, Selman B, Shah M (1997) Referral Web: Combining social networks and collaborative filtering. COMMUN ACM 40: 63–65. doi: 10.1145/245108.245123

4. Hannon J, Bennett M, Smyth B (2010) Recommending twitter users to follow using content and collaborative filtering approaches. Proc ACM RECSYS 2010: 199–206. doi: 10.1145/1864708.1864746

5. Morita M, Shinoda Y (1994) Information filtering based on user behavior analysis and best match text retrieval. Proc ACM SIGIR 1994: 272–281. doi: 10.1007/978-1-4471-2099-5_28

6. Claypool M, Le P, Wased M, Brown D (2001) Implicit interest indicators. Proc IUI 2001: 33–40. doi: 10.1145/359784.359836

7. Schein AI, Popescul A, Ungar LH, Pennock DM (2002) Methods and metrics for cold-start recommendations. Proc ACM SIGIR 2002: 253–260. doi: 10.1145/564376.564421

8. Linden G, Smith B, York J (2003) Amazon. com recommendations: Item-to-item collaborative filtering. IEEE INTERNET COMPUT 7: 76–80. doi: 10.1109/mic.2003.1167344

9. Huang Z, Zeng DD, Chen H (2007) Analyzing consumer-product graphs: Empirical findings and applications in recommender systems. MANAGE SCI 53: 1146–1164. doi: 10.1287/mnsc.1060.0619

10. Huang Z, Lin DKJ (2009) The time-series link prediction problem with applications in communication surveillance. INFORMS J COMPUT 21: 286–303. doi: 10.1287/ijoc.1080.0292

11. Liu H, Maes P (2005) Interestmap: Harvesting social network profiles for recommendations. Proc IUI 2005: 54–59.

12. Golbeck J (2006) Generating predictive movie recommendations from trust in social networks. INT FED INFO PROC 3986: 93–104. doi: 10.1007/11755593_8

13. Liu F, Lee HJ (2010) Use of social network information to enhance collaborative filtering performance. EXPERT SYST APPL 37: 4772–4778. doi: 10.1016/j.eswa.2009.12.061

14. Barabasi AL, Crandall R (2003) Linked: The new science of networks. AM J PHYS 71: 409. doi: 10.1119/1.1538577

15. Su X, Khoshgoftaar TM (2009) A survey of collaborative filtering techniques. LECT NOTES ARTIF INT 2009: 4. doi: 10.1155/2009/421425

16. Goldberg D, Nichols D, Oki BM, Terry D (1992) Using collaborative filtering to weave an information tapestry. COMMUN ACM 35: 61–70. doi: 10.1145/138859.138867

17. Resnick P, Iacovou N, Suchak M, Bergstrom P, Riedl J (1994) GroupLens: An open architecture for collaborative filtering of netnews. Proc ACM CSCW 1994: 175–186. doi: 10.1145/192844.192905

18. Breese JS, Heckerman D, Kadie C (1998) Empirical analysis of predictive algorithms for collaborative filtering. Morgan Kaufmann Publishers Inc. 43–52p.

19. Adomavicius G, Tuzhilin A (2005) Toward the next generation of recommender systems: A survey of the state-of-the-art and possible extensions. IEEE T KNOWL DATA EN 17: 734–749. doi: 10.1109/tkde.2005.99

20. Koren Y (2008) Factorization meets the neighborhood: A multifaceted collaborative filtering model. Proc ACM SIGKDD 2008: 426–434. doi: 10.1145/1401890.1401944

21. Su X, Khoshgoftaar TM (2006) Collaborative filtering for multi-class data using belief nets algorithms. Proc IEEE TAI 2006: 497–504. doi: 10.1109/ictai.2006.41

22. Chee S, Han J, Wang K (2001) RecTree: An efficient collaborative filtering method. LECT NOTES COMPUT SC 2114: 141–151. doi: 10.1007/3-540-44801-2_15

23. Ungar LH, Foster DP (1998) Clustering methods for collaborative filtering. Workshop on receommender systems at the 15th national conference on artificial intelligence 1998: 114–129.

24. Hofmann T, Puzicha J (1999) Latent class models for collaborative filtering. Proc IJCAI 1999: 688–693.

25. Huang Z, Zeng DD (2011) Why does collaborative filtering work? Transaction-based recommendation model validation and selection by analyzing bipartite random graphs. INFORMS J COMPUT 23: 138–152. doi: 10.1287/ijoc.1100.0385

26. Ho CW, Ruehli A, Brennan P (1975) The modified nodal approach to network analysis. IEEE T CIRCUITS SYST 22: 504–509. doi: 10.1109/tcs.1975.1084079

27. Lee JS, Jun CH, Lee J, Kim S (2005) Classification-based collaborative filtering using market basket data. EXPERT SYST APPL 29: 700–704. doi: 10.1016/j.eswa.2005.04.037

28. Chib S, Greenberg E (1995) Understanding the metropolis-hastings algorithm. AM STAT 49: 327–335. doi: 10.1080/00031305.1995.10476177

29. Melville P, Mooney RJ, Nagarajan R (2002) Content-boosted collaborative filtering for improved recommendations. Proc Natl AI 2002: 87–189.

30. Sarwar B, Karypis G, Konstan J, Riedl J (2000) Analysis of recommendation algorithms for e-commerce. Proc ACM ELECTRON COMMER 2000: 158–167. doi: 10.1145/352871.352887

31. Mild A, Reutterer T (2001) Collaborative filtering methods for binary market basket data analysis. Active Media Technology 2252: 302–313. doi: 10.1007/3-540-45336-9_35

32. Rousseeuw PJ, Kaufman L (1990) Finding Groups in Data. An Introduction to Cluster Analysis. Wiley Online Library.

33. Lemire D (2005) Scale and translation invariant collaborative filtering systems. INFORM RETRIEVAL 8: 129–150. doi: 10.1023/b:inrt.0000048492.50961.a6

34. Cechinel C, Sicilia MÁ, Sánchez-Alonso S, García-Barriocanal E (2012) Evaluating collaborative filtering recommendations inside large learning object repositories. INFORM PROCESS MANAG In press.

35. Bastian M, Heymann S, Jacomy M (2009) Gephi: An open source software for exploring and manipulating networks.

Chapter 5

ELECTRICAL CHARACTERIZATION OF TRAPS IN ALGAN/GAN FAT-HEMT'S ON SILICON SUBSTRATE BY C-V AND DLTS MEASUREMENTS

Manel Charfeddine[1], Malek Gassoumi[1], Hana Mosbahi[1], Christophe Gaquiére[2], Mohamed Ali Zaidi[1], Hassen Maaref[1]

[1]Laboratoire des Micro-Optoélectroniques et Nanostructures, Université de Monastir, Faculté des Sciences de Monastir, Monastir, Tunisie

[2]Institut d'Electronique de Microélectronique et de Nanotechnologie IMEN, Département Hyperfréquences et Semiconducteurs, Université des Sciences et Technologies de Lille, Villeneuve d'Ascq Cedex, France

ABSTRACT

We investigate high electron mobility transistors (HEMT's) based on AlGaN/GaN grown by molecular beam epitaxy on Silicon substrates. The improvement of the performances of such transistors is still subject to the influence of threading dislocations and point defects which are commonly observed in these devices. Deep levels in FAT-HEMT's are characterized by using Capacitance-Voltage (C-V) measurements, from which we can extract the barrier height and the donor concentration in the AlGaN layer. Deep Level Transient Spectroscopy (DLTS) Technique is also employed to identify defects in the heterostructure. Measurements reveal the presence of one electron trap with the activation energy $E_1 = 0.30$ eV and capture cross-section $\sigma_n = 3.59 \times 10^{-19}$ cm^2. The localization and the identification of this trap have been discussed.

INTRODUCTION

The material semiconductor GaN nitride and associated alloys (III-N compounds) arouse since more than three decades a regained interest because of their properties and their exceptional robustness [1-3]. These large gap materials, which extend from 0.7 to 6.1 eV, allow the realization of heterostructures in order of applications in the field of the ultra high frequencies (RF) power components [4,5]. A new market thus emerges in the field of telecommunications with High Electronic Mobility Transistors (HEMT's) which can support power densities 10 times higher than those accessible with the silicon and gallium arsenide technologies as reported by Dumka et al. [6], as well as cut-off frequencies higher than 100 GHz were reported in a previous paper [7]. These HEMT's structures are essential to meet the increasing needs for the communication systems requiring high power and frequency applications [8-10] (radars, stations of bases, connection satellite). Due to their strong thermal conductivity and good performance stability in a hostile environment, GaN-based HEMT devices are excellent candidates especially in high power/frequencies domain. Thus, many applications turn to the GaN-based components, in addition to RF power amplifiers [11,12].

Silicon substrate presents a significant attraction from its very competitive price, the size of the substrates available (up to 12"). It has a good thermal conductivity close to the one of GaN. In addition, the (111) orientation with a 6-fold symmetry is preferred for the GaNbased heterostructures on the silicon substrate [13]. The development of GaN-based devices on silicon mostly relies on the control of material quality, especially the management of the dislocation density and of the stress due to the thermal expansion coefficient mismatch between GaN and silicon. Thus, Silicon presents a certain number of determining advantages for the mass production of III-Nitride devices. The realization of powerful and reliable FAT-HEMT transistors III-N requires a buffer layer with good electric insulation proprieties, in addition to an electrical substrate insulator with a good thermal conductivity [14-16]. In FAT-HEMT (Al,Ga)N/ GaN heterostructures. The compensation of the residual doping of type N of the buffer layers is generally ensured by intrinsic mechanisms related to defects.

Although the significant progress obtained on the processes of development, few studies were devoted to the electric defect characterization, and the mechanisms of electric conduction in the devices. To this end, research works have been undertaken in the subject of trapping effects with the use of different electrical characterization techniques, including photoionization spectroscopy [17], drain leakage-current measurements [18], current-deep level transient spectroscopy (CDLTS) [19] and deep level transient spectroscopy (DLTS) [20], Current-Voltage, I(V) [21] and Capacitance-Voltage C(V) measurements [22].

In this paper, we intend to analyze the C(V) characteristics of FAT-HEMT AlGaN/GaN transistors based on Si substrate. Then we will characterize the electrical active defects present in the structure, using DLTS. This technique has been proved to be a powerful tool to probe the electronic properties of FAT-HEMT AlGaN/GaN transistors. The localization and the identification of these traps are discussed.

SAMPLE DESCRIPTION

The investigated structure consists of an AlGaN/GaN FAT-HEMT's grown on resistive silicon (111) substrate (4000 - 10.000 Ω cm) using ammonia (Riber Compact 21). The epitaxial layers are obtained by molecular beam epitaxy (MBE). Details on the epitaxial growth are reported elsewhere [23] and layer ticknesses are taken as follows: a 100 nm AlN nucleation layer is at first grown on the Si (111) substrate followed by a 2 μm thick GaN and a 30 nm-thick undoped AlGaN layer encapsulated with 1nm of an unintentionally doped (UID) GaN cap layer. This cap layer is grown to ensure the ohmic contact. The ohmic metallization contacts were prepared by evaporating Ti/Al/Ni/Au multilayer and rapid-thermal annealing at 850°C in a N_2 atmosphere for 30 s. For the gate contact, a thick Ni layer covered by Au layer (Ni/Au) and patterned by e-beam lithography is employed with a Schottky area of 250×200 μm^2.

Deep level transient spectroscopy (DLTS) is one of the most versatile techniques used to determine the electrical properties of defects. For that, it has been used as a technique to characterize the electron traps in the AlGaN/GaN/Si heterostructures. Measurements were performed using double lock-in detection and a PAR 410 capacitance meter with a frequency of 1 MHz recorded in the temperature range 10 - 320 K using Helium cooled cryostat.

EXPERIMENTAL RESULTS AND DISCUSSION

We first briefly discuss the C-V results. In order to validate the presence of a two-dimensional electron gas (2-DEG), we took measurements of Capacitance-Voltage (C-V) using a capacitance meter of 1MHz frequency (Figure 1). We note that the capacitance variation as a function of the gate voltage shows the existence of a capacitance plateau appearing from 8 V to 2 V, and it is associated to the depletion of the two-dimensional electron gas (2-DEG), located at the heterointerface. Between the plateau and the right part, there is a transition region, where the capacitance decreases rapidly with decreasing applied voltage. Then, we observe a sharp fall of the capacitance at the bias known as the pinch-off voltage ($V_p = 0.5$ V). The region beyond the V_p voltage is named the residual capacitance region. This is a characteristic of the 2-DEG. C(V) measurements allow estimating the values of the applied pulsate with voltage and the reverse bias in DLTS measurements.

The C-V characteristics of the AlGaN/GaN/Si FATHEMT's measured at different temperatures are presented in Figure 2. It shows good reproducibility even when a forward voltage of 0.5 V is applied.

This figure reveals that the capacitance decreases when increasing the temperature. The maximal value of the capacitance ranges from 0.82 to 1.1 nF. When the temperature increases, the pinch-off voltage position shifts towards the high voltage. The pinch-off voltage shift at 300 K is more gradual than that at 100 K. This significant shift informs us about the presence of the defects which are probably localized at the interface and in the vicinity of the accumulation region. The defects are present probably at the same range of temperatures. This will be confirmed by using DLTS technique.

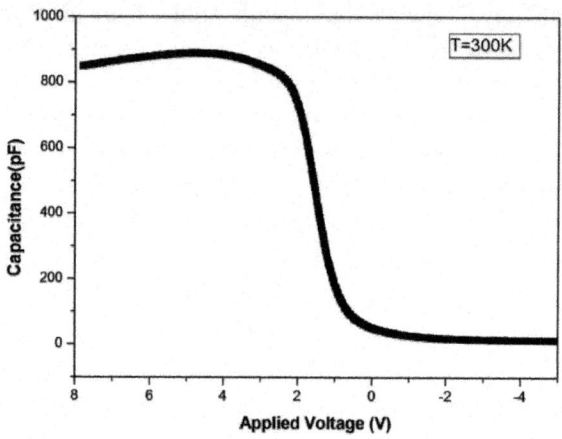

Figure 1: Capacitance-Voltage measurements results obtained on AlGaN/GaN/Si FAT-HEMTs at T = 300 K.

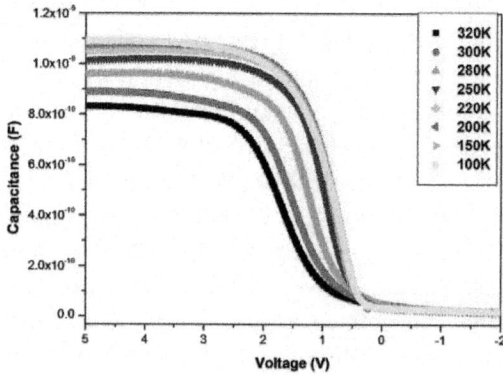

Figure 2: The C-V characteristics of the AlGaN/GaN/Si FAT-HEMTs measured at temperatures ranging from 100 K to 320 K.

We have deduced the carrier concentration profile N_{c-v} versus the space charge depth W in the heterostructure according to the following relation:

$$N_{C-V} = \frac{C^3}{q \varepsilon S^2 (dC/dV)}$$

(1)

and

$$W = S \frac{\varepsilon_0 \varepsilon_r}{C}$$

where S is the surface of the Schottky contact ε_0 is the free-space dielectric constant ε_r is the relative dielectric constant of AlGaN barrier and q is the elementary charge. The results are plotted in Figure 3. It exhibits a strong peak of a carrier density equals to 1.60×10^{21} cm^{-3} corresponding to the presence of a gas (2DEG) at the interface AlGaN/GaN. The position of this peak proves that AlGaN layer thickness is about 32 nm. In addition, we have determined the net doping concentration N_D and the barrier height φ_B from the lot of $1/C^2$ as a function of gate voltage (Figure 4). We have found $N_D = 4.4010^{19}$ cm^{-3} and $\varphi_B = 0.9$ eV.

To detect, locate and identify the localization of deep traps in FAT-HEMT transistors, we have carried out deep levels transient spectroscopy measurements for a reverse applied voltage of 2 V with an amplitude of 2 V, a filling time (t_p) was fixed to 0.5 ms and for different emission rate.

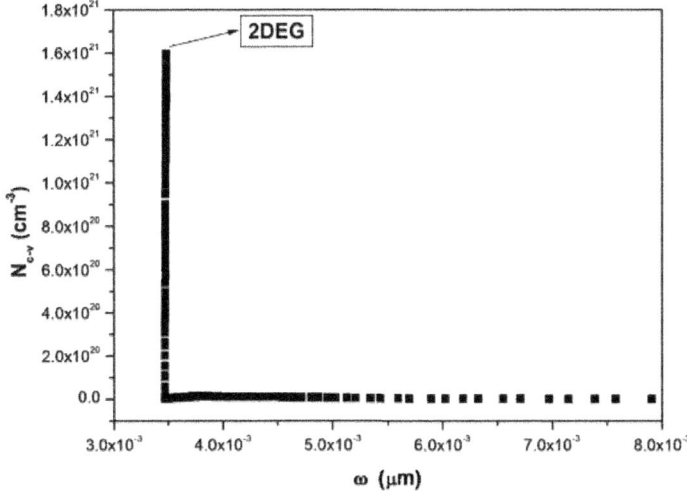

Figure 3: Carrier distribution profile.

Typical DLTS capacitance spectra (Figure 5) of AlGaN/GaN/Si HEMTs at V_{gs} = –2 V reveal the presence of only one peak corresponding to electron emission from one trap named E_1, The apparent activation energy E_a and the capture cross-section σ_n associated to the observed electron trap are extracted from the Arrhenius plot of: Ln (T^2/e_n) versus 1000/T .

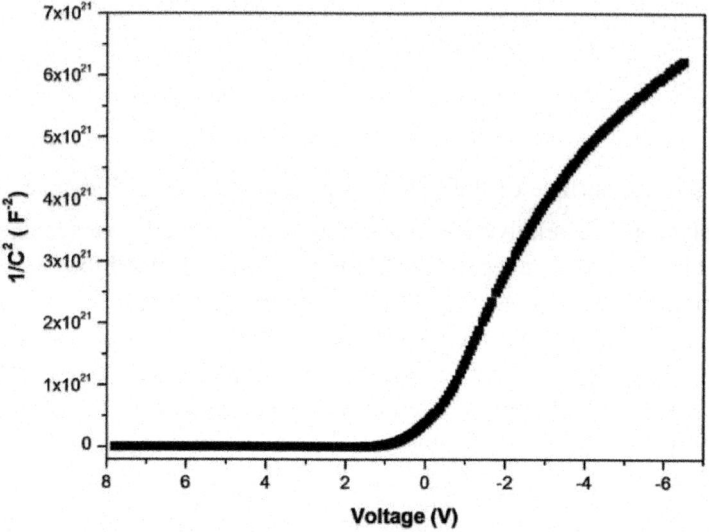

Figure 4: Curves represent $1/C^2$ according to Voltage.

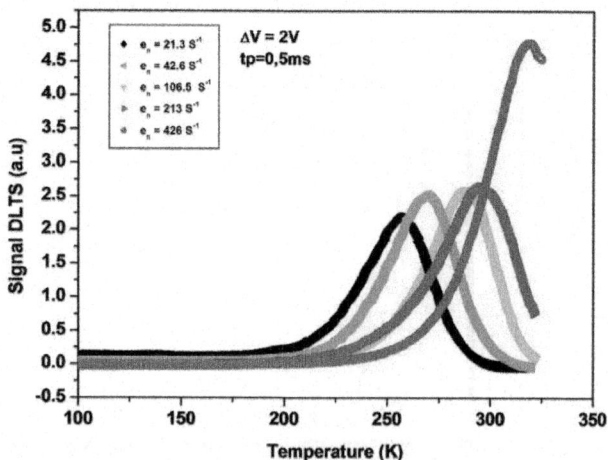

Figure 5: A typical capacitance DLTS spectrum showing the presence of one level performed on AlGaN/GaN/Si FAT-HEMTs. Recording condition: different emission rate, reverse bias 2 V, and filling time t_p = 0.5 ms.

The electrons are emitted as a consequence of thermal activation. The thermal emission rate e_n is written as [24]:

$$e_n = \delta_n T^2 \sigma \exp\left(\frac{-E_a}{KT}\right)$$

(2)

with

$$\delta_n = \frac{4\Pi\sqrt{6}\Pi^{3/2}K_\beta^2 m_n^*}{h^3}$$

This relation applied to the peak temperatures at which the electron emission rate, e_n, equals the rate window setting. In Equation (2), where $m_n^* = m_0 m_{GaN}$ such as m_n^* is the electron effective mass, m_0 is the mass of the free electron and $m_{GaN} = 0.22$ is the relative electrons mass in the GaN case, E_a the activation energy, σ_n the capture cross section, the temperature, h and k_β are the Planck and Boltzmann constant respectively. The capture cross section was assumed to be independent of temperature and the value of m_n^*. This Arrhenius plot allows an identification of a deep level defect. The results are presented in Figure 6.

The electron trap E_1, peaked at T = 318 K has an activation energy E_a = (0.30 ± 0.02) eV and a capture cross-section $\sigma_n = 3.59 \times 10^{-19}$cm². This E_a value was previously found by Sghaier et al. [25] observed a similar defect with an activation energy close to 0.31 eV by DLTS using boxcar technique. Nozaki et al. [26] have shown a comparable defect with an activation energy of 0.28 eV using I-DLTS (current-DLTS) technique performed on AlGaN/GaN MODFET. They attribute this defect to dislocation localized in the AlGaN layer. This defect was also studied by Tang [27], with a sample growth of unintentionally doped (UID) semi-insulating GaN on SiC which is mainly employed in GaN HEMT devices. To investigate the defect proprieties and their signatures, they used the PL technique correlated with a typical net TSC electric technique. They obtained an activation energy E_a = 0.30 eV. This value is in good agreement with the value characterizing the defect found in our heterostructure. The deep level, as an extendeddefect, has been also widely observed by DLTS in n-type GaN and AlGaN/GaN heterostructures [28].

The trap carrier concentration is extracted according to the relation (3) with the approximation $N_T < N_D$:

$$\frac{\Delta C}{C_0} \approx \frac{1}{2}\frac{N_T}{N_D}$$

(3)

The carrier concentration obtained is about $N_T = 6.85 \times 10^{16}$ cm^{-3} in our sample Si doped. Nevertheless, it should be mentioned that the trap with electrons located at approximately 0.30 eV of the conduction band is present into strong concentration probably due to Si dopage. A similar carrier concentration of this defect is observed by

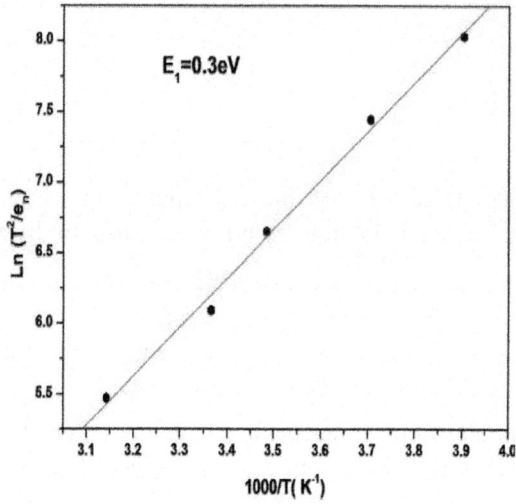

Figure 6: Arrhenius plots for E$_1$ deep levels observed in AlGaN/GaN/Si FAT-HEMTs.

Mohamed et al. [29] by using charge deep level transient spectroscopy (Q-DLTS) in the n-GaN sample. Our results seem to be a dislocation extended from buffer layer to the AlGaN layer.

CONCLUSIONS

In summary, we have investigated static measurements and defect analysis on FAT-HEMT AlGaN/GaN realized on Si substrate grown by MBE. Defects analysis performed on these transistors by C-V characteristics and DLTS Signal shows one deep defect that we extracted its signature using on the Arrhenius plot. This trap is attributed to dislocation localized in the AlGaN layer. Finally, the extraction of the deep trap concentration has been evidenced for AlGaN/GaN FAT-HEMTs.

REFERENCES

1.   P. Omling, E. R. Weber, L. Montelius, H. Alexander and J. Michel, "Electrical Properties of Dislocations and Point Defects in Plastically Deformed Silicon," Physical Review B, Vol. 32, No. 20, 1985, pp. 6571-

6581. doi:10.1103/PhysRevB.32.6571

2. M. A. Khan, A. Bhattarai, J. N. Kuznia and D. T. Olson, "High Electron Mobility Transistor Based on a GaN-Al$_x$, Ga$_{1-x}$N Heterojunction," Applied Physics Letters, Vol. 63, No. 9, 1993, pp. 1214-1215. doi:10.1063/1.109775

3. L. Semra, A. Teliaa and A. Soltani, "Trap Characterization in AlGaN/GaN HEMT by Analyzing Frequency Dispersion in Capacitance and Conductance," Surface and Interface Analysis, Vol.42, 2010, pp. 799-802. doi:10.1002/sia.3462

4. H. Kim, S. J. Park and H. Hwang, "Design and Fabrication of Highly Efficient GaN-Based Light-Emitting Diodes," IEEE, Vol. 49, No. 10, 2002, pp. 1715-1721.

5. R. D. Dupuis, "Epitaxial Growth of III-V Nitride Semiconductors by Metalorganic Chemical Vapor Deposition," Journal of Crystal Growth, Vol. 178, 1997, pp. 56-73.doi:10.1016/S0022-0248(97)00079-1

6. D. C. Dumka, C. Lee, H. Q. Tserng, P. Saunier and M. Kumar, "AlGaN/GaN HEMTs on Si Substrate with 7 W/mm Output Power Density at 10 GHz," Electronics Letters, Vol.40, 2004, pp. 1023-1024. doi:10.1049/el:20045292

7. A. Minko, V. Hoël, E. Morvan, B. Grimbert, A. Soltani, E. Delos, D. Ducatteau, C.Gaquière, D. Théron, J. C. D. Jaeger, H. Lahreche, L. Wedzikowski, R. Langer and P. Bove, "AlGaN-GaN HEMTs on Si with Power Density Performance of 1.9 W/mm at 10 GHz," IEEE Electron Device Letters, Vol. 25, 2004, pp. 453-455. doi:10.1109/LED.2004.830272

8. M. A. Khan, J. N. Kuznia, D. T. Olson, J. M. Van Hove, M. Blasingame and L. F. Reitz, "High-Responsivity Photo-Conductive Ultraviolet Sensors Based on Insulating Single-Crystal GaN Epilayers," Applied Physics Letters, Vol. 60, 1992, pp. 2917-2919.doi:10.1063/1.106819

9. S. Sttrite and H. Morkoç, "GaN, AlN, and InN: A Review," Journal of Vacuum Science & Technology B, Vol. 10, 1992, pp. 1237-1266. doi:10.1116/1.585897

10. M. A. Khan, J. N. Kuznia, D. T. Olson, M. Blasingame and A. R. Bhattarai, "Schottky Barrier Photodetector Based on Mg-Doped p-Type GaN Films," Applied Physics Letters, Vol.63, 1993,pp. 2455-2456. doi:10.1063/1.110473

11. X. L. Wang, T. S. Shen, H. L. Xiao, C. M. Wang, G. X. Hu, W. J. Luo, J. Tang, L. C. Guo and J. M. Li, "HighPerformance 2 mm Gate Width GaN HEMTs on 6H-SiC with Output Power of 22.4 W@ 8 GHz," Solid-State

Electron, Vol. 52, 2008, pp. 926-929.doi:10.1016/j.sse.2007.12.014

12.   D. Ducatteau, A. Minko, V. Hoël, E. Morvan, E. Delos, B. Grimbert, H. Lahreche, P. Bove, C. Gaquière, J. C. De Jaeger and S. Delage, "Output Power Density of 5.1/mm at 18 GHz with an AlGaN/GaN HEMT on Si Substrate," IEEE Electron Device Letters, Vol. 27, 2006, pp. 7-9. doi:10.1109/LED.2005.860385

13.   T. Ito, Y. Nomura, S. L. Selvaraj and T. Egawa," Comparison of Electrical Properties in GaN Grown on Si(111) and c-Sapphire Substrate by MOVPE," Journal of Crystal Growth, Vol. 310, 2008, pp. 4896-4899. doi:10.1016/j.jcrysgro.2008.08.029

14.   N. Maeda, K. Tsubaki, T. Saitoh, T. Tawara and N. Kobayashi, "High-Temperature Electron Transport Properties in AlGaN/GaN Heterostructures," Applied Physics Letters, Vol. 79, 2001, pp. 1634-1636. doi:10.1063/1.1400779

15.   N. Maeda, T. Saitoh, K. Tsubaki, T. Nishida and N. Kobayashi, "Two-Dimensional Electron Gas Transport Properties in AlGaN/GaN Singleand Double-Heterostructure Field Effect Transistors," Materials Science and Engineering: B, Vol. 82, 2001, pp. 232-237.doi:10.1016/S0921-5107(01)00604-3

16.   N. Maeda, T. Nishida, N. Kobayashi and M. Tomizawa, "Two-Dimensional Electron-Gas Density in $Al_xGa_{1-x}N$/ GaN Heterostructure Field-Effect Transistors," Applied Physics Letters, Vol. 73, No. 13, 1998, pp. 1856-1858. doi:10.1063/1.122305

17.   P. B. Klein, "Photoionization Spectroscopy in AlGaN/ GaN High Electron Mobility Transistors," Journal of Applied Physics, Vol. 92, 2002, pp. 5498-5502.doi:10.1063/1.1510564

18.   S. Arulkumaran, T. Egawa, H. Ishikawa and T. Jimbo, "Comparative Study of Drain-Current Collapse in AlGaN/GaN High-Electron-Mobility Transistors on Sapphire and Semi-Insulating SiC," Applied Physics Letters, Vol. 81, No. 16, 2002, pp. 3073-3075.doi:10.1063/1.1512820

19.   O. Mitrofanov, M. Manfra and N. Weimann, "Impact of Si Doping on Radio Frequency Dispersion in Unpassivated GaN/AlGaN/GaN High-Electron-Mobility Transistors Grown by Plasma-Assisted Molecular-Beam Epitaxy," Applied Physics Letters, Vol. 82, No. 24, 2003, pp. 4361-4363. doi:10.1063/1.1582373

20.   A. P. Zhang, L. B. Rowland, E. B. Kaminsky, V. Tilak, J. C. Grande, J. Teetsov, A. Vertiatchikh and L. F. Eastman, "Correlation of Device Performance and Defects in AlGaN/GaN High-Electron Mobility Transistors," Journal of Electronic Materials, Vol. 32, No. 5, 2003, pp.

388- 394. doi:10.1007/s11664-003-0163-6

21. M. Gassoumi, J. M. Bluet, G. Guillot, C. Gaquière and H. Maaref, "Characterization of Deep Levels in High Electron Mobility Transistor by Conductance Deep Level Transient Spectroscopy," Materials Science and Engineering C, Vol. 28, 2008, pp. 787-790.doi:10.1016/j. msec.2007.10.068

22. R. Mosca, E. Gombia, A. Passaseo, V. Tasco, M. Peroni and P. Romanini, "DLTS Characterization of Silicon Nitride Passivated AlGaN/GaN Heterostructures," Superlattices and Microstructures, Vol. 36, 2004, pp. 425-433. doi:10.1016/j.spmi.2004.09.006

23. V. Hoel, N. Vellas, C. Gaquiere, J. C. DeJaeger, Y. Cordier, F. Semond, F. Natali and J. Massies, "High-Power AlGaN/GaN HEMTs on Resistive Silicon Substrate," Electronics Letters, Vol. 38, 2002, pp. 750-752. doi:10.1049/el:20020522

24. J. Osaka, Y. Ohno, S. Kishimoto, K. Maezawa and T. Mizutani, "Deep Levels in n-Type AlGaN Grown by Hydride Vapor-Phase Epitaxy on Sapphire Characterized by Deep-Level Transient Spectroscopy," Applied Physics Letters, Vol. 87, 2005, pp. 222112-1-222112-3. doi:10.1063/1.2137901

25. N. Sghaier , M. Trabelsi , N. Yacoubi, J. M. Bluet, A. Souifi, G. Guillot, C. Gaquière and J. C. Dejaeger, "Traps Centers and Deep Defects Contribution in Current Instabilities for AlGaN/GaN HEMT's on Silicon and Sapphire Substrates," Microelectronics Journal, Vol. 37, 2006, pp. 363-370. doi:10.1016/j.mejo.2005.05.014

26. S. Nozaki, H. Feick, E. R. Weber, M. Micovic and C. Nguyen, "Compression of the dc Drain Current by Electron Trapping in AlGaN/ GaN Modulation Doped FieldEffect Transistors," Applied Physics Letters, Vol. 78, No. 19, 2001, pp. 2896-2898. doi:10.1063/1.1367274

27. H. Tang, Z. Q. Fang, S. Rolfe, J. A. Bardwell and S. Raymond, "Growth Kinetics and Electronic Properties of Unintentionally Doped Semi-Insulating GaN on SiC and High-Resistivity GaN on Sapphire Grown by Ammonia Molecular-Beam Epitaxy," Journal of Applied Physics, Vol. 107, 2010, pp. 103701-1-103701-12.

28. Z.-Q. Fang, D. C. Look, D. H. Kim and I. Adesida, "Traps in AlGaN/GaN/ SiC Heterostructures Studied by Deep Level Transient Spectroscopy," Applied Physics Letters, Vol. 87, 2005, pp. 182115-1-182115-3. doi:10.1063/1.2126145

29. Z. H. Mahmood, A. P. Shah, A. Kadir, M. R.Gokhale, A. Bhattachary and B. M. Arora, "Charge Deep Level Transient Spectroscopy of Electron Traps in MOVPE Grown n-GaN on Sapphire," Physica Status Solidi, Vol. 245, 2008, pp. 2567-2571.

Chapter 6

RELIABILITY MEASUREMENT FOR MIXED MODE FAILURES OF 33/11 KILOVOLT ELECTRIC POWER DISTRIBUTION STATIONS

Faris M. Alwan[1, 3], Adam Baharum[1], Geehan S. Hassan[2, 4]

[1]School of Mathematical Sciences, University Sains Malaysia, Penang, Malaysia

[2]School of Computer Sciences, University Sains Malaysia, Penang, Malaysia

[3]Statistics Department, College of Administration and Economics, Baghdad University, Baghdad, Iraq

[4]Ibn-Rushud College of Education, Baghdad University, Baghdad, Iraq

ABSTRACT

The reliability of the electrical distribution system is a contemporary research field due to diverse applications of electricity in everyday life and diverse industries. However a few research papers exist in literature. This paper proposes a methodology for assessing the reliability of 33/11 Kilovolt high-power stations based on average time between failures. The objective of this paper is to find the optimal fit for the failure data via time between failures. We determine the parameter estimation for all components of the station. We also estimate the reliability value of each component and the reliability value of the system as a whole. The best fitting distribution for the time between failures is a three parameter Dagum distribution with a scale parameter $\beta = 90.001$ and shape parameters $k = 0.33998$ and $\alpha = 2.4011$. Our analysis reveals that the reliability value decreased by 38.2% in each 30 days. We believe that the current paper is the first to address this issue and its analysis. Thus, the results obtained in this research reflect its originality. We also suggest the practicality of using these results for power systems for both the maintenance of power systems models and preventive maintenance models.

RELIABILITY AND FAILURE FUNCTIONS

Electric power is transmitted through an electric circuit. The terms "high voltage" and "high power" indicate that the voltages of the electrical energy are high enough to inflict harm or death on living things. Electrical power systems are highly complex and extremely integrated. Reliability is one of the most important factors considered in the planning, design, operation, and maintenance of electric power systems [1], [2]. This factor is one of the most effective indicators of product quality that buyers take into account when choosing among different varieties [3]. Moreover, reliability generally becomes more important to consumers, as failure, repair, and maintenance entail expensive costs [3]. The reliability function is a mathematical and engineering indicator that is used to describe the state of the equipment in the system through the probability function. Many factors and definitions are related to reliability (e.g., mean time to failure [MTTF], mean time between failures [MTBF], and mean time to repair [MTTR]). The MTTF is the expected value representing the return period of equipment failure[4]–[7]. It can be expressed mathematically as [8], $\mathrm{MTTF} = E(T) = \int_{-\infty}^{\infty} t f(t) dt$, where $E(t)$ is the expected value of time, and $f(t)$ is the probability density function (pdf) for variable t. The MTTF is also referred to as the expected life. The mean time between failures (MTBF) and the MTTR are defined in Section Availability. The term *reliability* can be defined in many ways. For example, for an electrical switch, reliability may be defined as the probability that it successfully functions under a stipulated load and at a specific temperature. An operational definition of reliability must be sufficiently precise to establish a clear distinction between reliable and unreliable items. In addition, this definition must be sufficiently general to account for the complexities that arise in making this determination [8]. Based on this definition of reliability, reliability analyses often involve the analysis of binary outcomes *(0, 1)* (i.e., success=1, failure data=0) [8].

Assume that the period of failure T is a continuous random variable with values in a positive real line. Many methods are available to specify the properties of a random variable [8]. The first method involves using the pdf, $f(t)$, that satisfies, $f(t) \geq 0$ and $\int_{-\infty}^{\infty} f(t) dt = 1$.

When T is a Dagum random variable, its *pdf* is [9]–[11]

$$f(t) = k\alpha \left(1 + \left(\frac{t}{\beta}\right)^{\alpha}\right)^{-1-k} \beta^{-k\alpha} t^{-1+k\alpha} \quad t > 0$$

(1)

in which $t > 0$, β is the scale parameter ($\beta > 0$), α (> 0) and k (> 0) are the shape parameters.

A second method to specify the properties of T is the *cumulative distribution function*. Mathematically, this function is expressed as [12].

$$F(t) = P(T \leq t) = \int_{-\infty}^{t} f(s)ds,$$

where $f(s)$ is a pdf. The *cumulative distribution function* is the complement of the *reliability function*, and thus, it is called the *unreliability function* [8]. The *cumulative distribution function* for a Dagum random variable is

$$F(t) = \left(1 + \left(\frac{t}{\beta}\right)^{-\alpha}\right)^{-k} t > 0.$$

(2)

A third method to specify the properties of a random variable is through its *reliability function*, also known as the *survival function* [8]. We define the *reliability function* as

$$R(t) = P(T > t) = \int_{t}^{\infty} f(s)ds,$$

where $f(s)$ is a pdf. The *reliability function* for a Dagum random variable is [9], [13], [14].

$$R(t) = \beta^{-k\alpha}\left(1 + \left(\frac{t}{\beta}\right)^{\alpha}\right)^{-k} t^{k\alpha} t > 0.$$

(3)

The fourth method specifies the properties of a random variable as the *hazard function*, also called the *instantaneous failure rate function* (further details are provided in [8]).

$$h(t) = \frac{f(t)}{R(t)}.$$

The *hazard function* for a Dagum random variable is [14]

$$h(t) = \frac{k\alpha}{\left(1 + \left(\frac{t}{\beta}\right)^{\alpha}\right)\left(-1 + \left(1 + \left(\frac{\beta}{t}\right)^{\alpha}\right)^{k}\right)t} \quad t > 0$$

(4)

The functions $f(t)$, $F(t)$, $R(t)$, and $h(t)$ are called "failure functions."

Availability

At first, we have to define several factors that are closely associated with availability (e.g., failure, availability, and so on). Failure is defined as the incapability of the system (subsystem or one of its components) to perform its job [15], [16] or the "inability of the item to meet the requirements of the work"[17], [18]. The term "available" is defined as the state of an item such that it can perform its function under stated conditions of use and maintenance in the required location [19]. Most researchers define availability as the probability that an item will be available [19], [20] or the probability that the system will operate satisfactorily at any point in time when operating under a specified condition [20], [21]. $\text{Availability} = \dfrac{UT}{UT+DT} = \dfrac{MTBF}{MTBF+MTTR}$

where UT is the uptime or operating time, DT is downtime (excluding free time), MTBF is the mean time between failures, and MTTR is the mean time to repair. For a more accurate quantity, "inherent availability" is defined as [19], [21]:

$$\text{Inherent availability} = \frac{UT}{UT+ART}$$

where ART is the active repair time. The MTBF is defined by Frankel, Dinesh and Bryant [15],[22], [23] as a parameter of basic reliability of the repairable components. It is the ratio of the total number of life units for components of the total number of failures. MTTR is the mean time to repair, it is defined as the whole time required to manage the failure, including factors: the way in which the fault is detected and the response speed of the maintenance team with the repair time [24]. The mean corrective maintenance time is defined by Dhillon [25] as the main criterion of the maintainability of repairable items. It represents the average (mean) time required to repair failed equipment. This criterion can be observed based on the inherent availability equation wherein availability is integrated between reliability and the times for maintenance and repair [26]. This relationship is very important because it is used to express the probability that the system will be operating according to the mission time without failure[26].

Description of the Problem

The electric power distribution station in Iraq was designed to have two power transformers (T1and T2). Each transformer has a circuit breaker with limited capacity (1,200 A), denoted as CBT*i* functions as a main circuit breaker of the transformer. These two transformers are connected to the communication bus situated between them (termed as Bus-Bar). The Bus-Bar feeds group of feeders (10 feeders). The Bus-Bar was divided into two parts separated by circuit breaker with a limited capacity of 800 A, called the Bus-Bar circuit breaker (CBB). Each one of the ten feeders has a circuit breaker with a limited capacity of 400 A, called as (CBF*i*). The main circuit breakers must be switched *ON* and the CBB must be switched *OFF* if the transformers are in normal operation. However, if one of these transformers fails, the CBT of the failed transformer must be switched *OFF*, and the CBB is switched *ON* to provide power to the failed transformer feeders.

Data Collection

Five years data of the electricity distribution company in Baghdad, Iraq, were collected for time between failures (TBFs) of the electric power distribution station. The failure data were recorded manually. To deal with this problem, we reported the number of breakdowns within five years, and also the time between them. This is shown in Table 1. The first column inTable 1 represents failures numbers, and second column mentioned to the number of days require for the station to step-down. For example, the first TBF value was calculated between 12 am on 1*st* January and 11:50 pm on 14*th* April. Accordingly, the first step-down was occurred after (104.895833).

Table 1: Time between failures of the electric power distribution station for five years

Failure No.	TBFs(day)	Failure No.	TBFs(day)
1	104.895833	2	22.8854167
3	36.729167	4	0.9791667
5	54.8854167	6	83.6875
7	50.895833	8	6.83333
9	97.83333	10	42.8854167
11	149.9791667	12	6.9791667
13	13.9375	14	2.9375
15	70.9791667	16	109.83333
17	36.83333	18	47.78125
19	2.9375	20	118.8020833
21	9.83333	22	529
23	30.895833	24	15.78125
25	71.8854167	26	38.7604167
27	57.8854167	-	-

doi:10.1371/journal.pone.0069716.t001

doi:10.1371/journal.pone.0069716.t001

Each component of the station can be failed in random manner. The component average time between failures was modeled to be random variable following certain distribution [2]. EasyFit is the distribution fitting software that can be used to fit the appropriate statistical distribution for the TBFs. In the next section we will focus on the goodness of fit to find the best fitting statistical distribution for each component of TBFs.

Goodness of Fit

In this paper a goodness of fit for the TBFs statistical distribution was tested. Such test can be done using many tools. EasyFit software was used to perform this task. It includes the using of the *Kolmogorov-Smirnov, Anderson-Darling* and *Chi-square* test. The idea behind the goodness of fit tests is to have the "distance," critical values, measured between the data and the distribution being tested. Then that critical value is compared to some threshold value. The goodness of fit reports includes the test statistics and the critical values calculated for various significance levels (δ=0.2, 0.1, 0.05, 0.02, 0.01). Furthermore, the goodness of fit test statistics indicates the distance between the data and the provided distributions [27]. The P-value can be helpful specifically when the null hypothesis is rejected at all selected significance levels, where the P-value is criteria uniformity between the results actually obtained in the experiment and the random chance explanation for those results [28]–[30]. It is required to know at which level it could be accepted [31]. EasyFit deals with data using histogram based on TBFs samples. The number of vertical bars was based on the total number of observations (27 values). The equation $Q = 1 + \log 2N$, was used to find the number of bins (histogram), where N is the total number of TBFs and Q is the resulting number of classes [32]. The height of each histogram bar indicates how many data points fall into that class. To obtain the best fitting model, we chose various distributions. Our analysis reveals that the distribution with the lowest statistical value is the best-fitting model.Similar conclusion is also drawn by Fakhraei [33], This support the validity of our analysis. Based on this fact, each distribution is ranked (1= the best model, 2= the next best model and so on). The data was analyzed and tested under several nonnegative distributions using the EasyFit software. Dagum distribution is the optimal analysis of the TBFs, with scale parameter β=90.001 and shape parameters k=0.33998 and α=2.4011. Table 2 shows the summary of the goodness of fit of TBFs for the (39) nonnegative distributions. Table 3 shows the goodness of fit details of TBFs for Dagum distribution. Figure 1 shows only the six much closer distributions from all 39 nonnegative distributions. Figure 2 shows the fitting result of TBFs histogram with the Dagum distribution while Figure 3 (a, b, c and d) shows the failure functions (Pdf, CDF, Reliability function and Hazard function) respectively, of Dagum distribution (k=0.33998, α=2.4011, β=90.001).

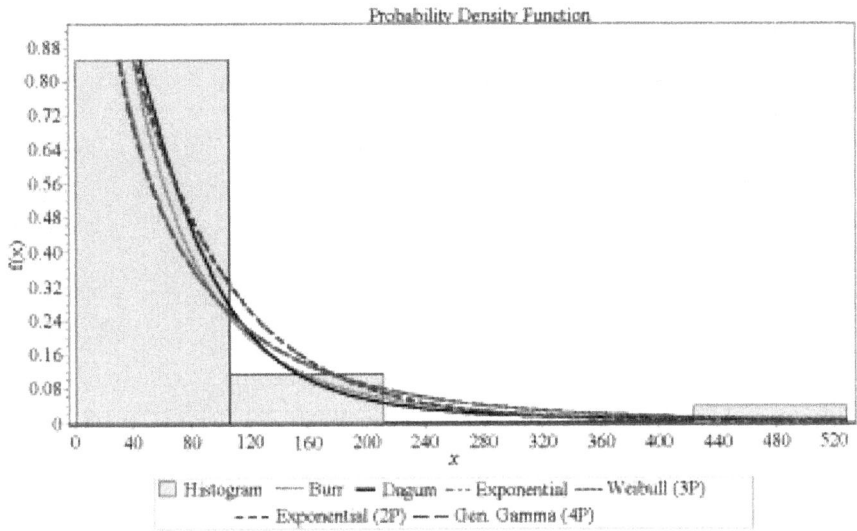

Figure 1: The fitting result for best six distributions of TBFs histogram.

doi:10.1371/journal.pone.0069716.g001

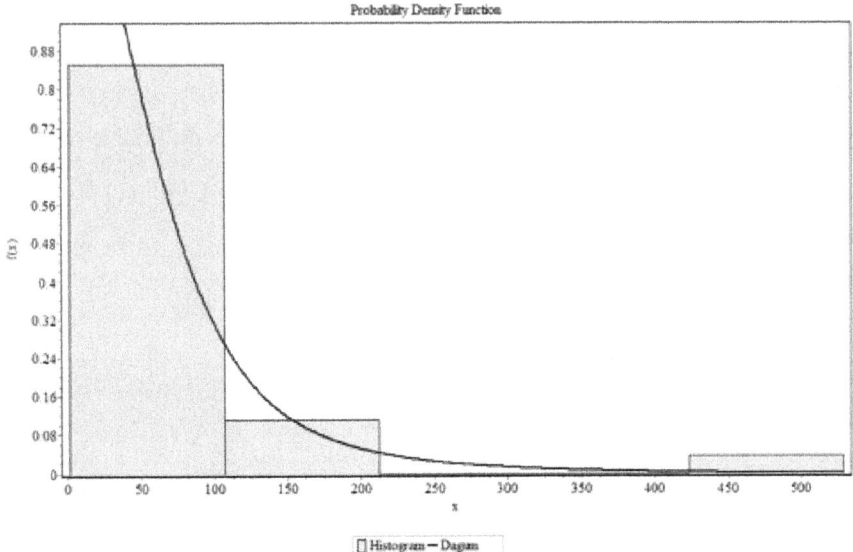

Figure 2: The Dagum distribution fitting result with TBFs data histogram.

doi:10.1371/journal.pone.0069716.g002

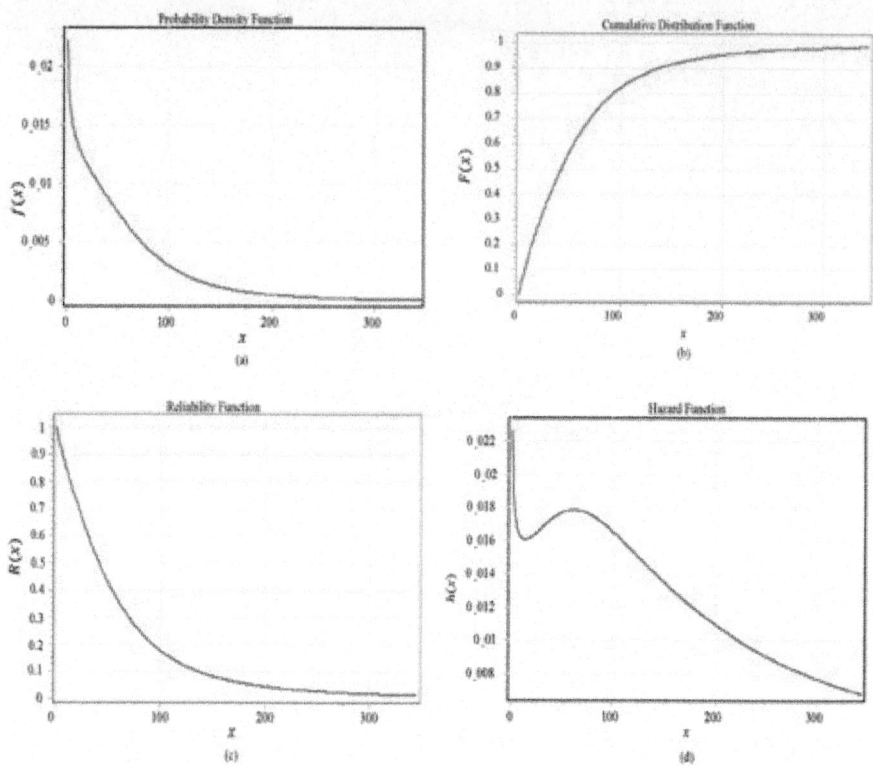

Figure 3: The failure functions of TBFs data of a Dagum random variable with

$k = 0.33998$, $a = 2.4011$ and $b = 90.001$, (a) Pdf, (b) CDF, (c) Reliability function and (d) Hazard function.

doi:10.1371/journal.pone.0069716.g003

Table 2: The summary of goodness of fit sorted by rank resulting from the*Kolmogorov-Smirnov* test

Distribution	Kolmogorov Smirnov Statistic	Rank	Anderson Darling Statistic	Rank	Chi-Squared Statistic	Rank
Dagum	0.09309	1	0.19582	1	0.49359	10
Exponential	0.09714	2	0.55542	11	1.2128	15
Exponential (2P)	0.09871	3	1.8547	22	0.78094	11
Weibull (3P)	0.11924	4	1.0625	15	0.24672	8
Burr	0.11966	5	0.3036	4	0.11459	4
Gen. Gamma (4P)	0.12057	6	1.1189	16	0.24378	7
Pearson 6	0.12139	7	0.30249	3	0.11222	3
Pareto 2	0.12214	8	0.29996	2	0.16934	5
Frechet (3P)	0.12715	9	0.3937	5	0.82819	12
Gamma (3P)	0.12922	10	4.2307	28	N/A*	
Log-Logistic (3P)	0.13296	11	0.51477	10	1.2346	17
Weibull	0.14339	12	0.41791	6	0.42538	9
Burr (4P)	0.1438	13	4.313	29	N/A*	
Inv. Gaussian (3P)	0.14394	14	0.44184	7	0.83618	14
Lognormal (3P)	0.1458	15	0.44902	8	1.2296	16
Fatigue Life (3P)	0.14646	16	0.47478	9	0.82997	13
Gen. Gamma	0.15678	17	0.76019	13	1.2862	19
Inv. Gaussian	0.16775	18	2.0065	24	1.2995	20
Lognormal	0.16859	19	0.63295	12	1.2509	18
Gamma	0.18948	20	1.5649	19	1.5286	21
Log-Logistic	0.19874	21	0.83928	14	2.1911	22
Pearson 6 (4P)	0.20203	22	4.7384	30	N/A*	
Fatigue Life	0.22105	23	1.299	17	3.883	26
Pearson 5 (3P)	0.22482	24	3.6623	27	0.00149	1
Levy	0.22647	25	1.8258	21	3.4804	24
Levy (2P)	0.22918	26	1.496	18	0.21343	6
Chi-Squared (2P)	0.24256	27	3.1159	25	4.8578	27
Frechet	0.2431	28	1.6912	20	0.01378	2
Rayleigh (2P)	0.24822	29	3.1436	26	3.6354	25
Pearson 5	0.25076	30	2.0032	23	2.3495	23
Rayleigh	0.25405	31	8.3926	32	7.157	28
Pareto	0.31571	32	6.5738	31	8.711	29
Rice	0.45759	33	18.714	33	31.134	30
Chi-Squared	0.5208	34	137.97	35	31.148	31
Dagum (4P)	0.52183	35	19.63	34	62.473	32
Erlang	No fit					
Erlang (3P)	No fit					
Log-Gamma	No fit					
Nakagami	No fit					

*: No answer.
doi:10.1371/journal.pone.0069716.t002

doi:10.1371/journal.pone.0069716.t002

Table 3: The details for goodness of fit for a Dagum distribution (3P)

Kolmogorov Smirnov					
Sample Size	27				
Statistic	0.09309				
P-Value	0.95633				
Rank	1				
δ	0.2	0.1	0.05	0.02	0.01
Critical Value	0.2003	0.22898	0.25438	0.28438	0.30502
Reject?	No	No	No	No	No
Anderson Darling					
Sample Size	27				
Statistic	0.19582				
Rank	1				
δ	0.2	0.1	0.05	0.02	0.01
Critical Value	1.3749	1.9286	2.5018	3.2892	3.9074
Reject?	No	No	No	No	No
Chi-Squared					
Deg. of freedom	3				
Statistic	0.49359				
P-Value	0.9203				
Rank	10				
δ	0.2	0.1	0.05	0.02	0.01
Critical Value	4.6416	6.2514	7.8147	9.8374	11.345
Reject?	No	No	No	No	No

doi:10.1371/journal.pone.0069716.t003

doi:10.1371/journal.pone.0069716.t003

MAXIMUM LIKELIHOOD ESTIMATION

The best result of goodness of fitting to the TBFs under many distributions using EasyFit software is the Dagum distribution. Alwan et al. [34] provides a more detailed treatment of the fitting method. The maximum likelihood method is used to estimate the parameters k, α, β of the Dagum distribution. The likelihood function, $L(\theta)$, from a generic distribution with density and reliability functions $f(.; \theta)$ and $R(.; \theta)$, respectively, can be written as [9].

$$L(\theta) = \prod_{i=1}^{n} f(t_i; \theta) R(t_i; \theta)$$

where θ is the parameter(s) of the distribution and $i = 1,2,...,n$. Consider a sample of size n (which is 27 samples in this paper). The log-likelihood function for the estimate of the parameters $\theta = (k, \alpha, \beta)$ of the Dagum distribution is given by Domma et al. [9]. That is, the log-likelihood function, $\ell(\theta)$, based on data from Eq.1 is [9].

$$\ell(\theta) = \sum_{i=1}^{n} \{\ln(k\alpha\beta) - (\alpha+1)\ln(t_i) - (k+1)\ln(1+\beta t_i^{-\alpha})\}$$

$$+ \sum_{i=1}^{n} \{\ln\{1 - [1+\beta t_i^{-\alpha}]^{-k}\}\}$$

(5)

The MLEs $\hat{\theta} = (\hat{k}, \hat{\alpha}, \hat{\beta})$ are obtained from the numerical maximization of Eq.5, since the solution of the maximum likelihood equations is not in closed form [9]. Using Eq.5 the values of the estimated k, α, β parameters for each component of the station relying on the maximum likelihood method are presented in Table 4.

Table 4: Estimated scale and shape parameters of Dagum distribution for each component

components	K	α	β
T_1	0.1319	4.0147	111.0092
T_2	0.1283	5.5402	163.794
CBT_1	20.8619	0.6425	0.110026
CBT_2	0.0764	12.2022	120.841
CBF_1	12.4569	1.18365	3.75407
CBF_2	2.02372	1.24288	13.0104
CBF_3	0.69703	1.45807	32.3148
CBF_4	0.929475	1.23822	21.3319
CBF_5	1.83028	1.20514	14.535
CBF_6	178.257	1.3475	0.403921
CBF_7	9.52767	0.3868	0.01772
CBF_8	1.37434	1.53844	37.4659
CBF_9	1.92766	1.45856	12.6204
CBF_{10}	1.50505	0.18518	14.0558

doi:10.1371/journal.pone.0069716.t004

doi:10.1371/journal.pone.0069716.t004

RELIABILITY ASSESSMENT

Figure 4 shows the reliability block diagram for the electric power distribution station. It also represent the visualization of the components working. The reliability function for a Dagum random variable was provided in Eq.3. In the section Problem Statement, the CBB does not function if, and only if, one of the transformers do not operate. Fourteen different components exist, excluding the CBB. Based on Figure 4, the following classifications of the block diagram reliability of the system for the electric power distribution station have been described as

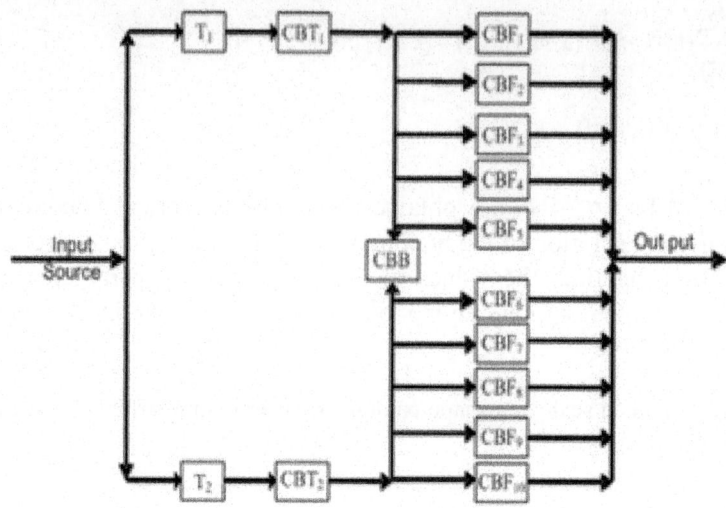

Figure 4: The reliability block diagram of the 33/11 KV electric power distribution station.

doi:10.1371/journal.pone.0069716.g004

First Group(FG)

Transformers 1 and 2 as well as the main circuit breaker are connected together in a series. At the same time, the two transformers (Transformers 1 and 2) and their main circuit breaker are connected in a parallel manner (see Figure 4). The reliability function of this group is expressed as

$$R_{FG}(t) = 1 - \prod_{i=1}^{2} (1 - (R_{Ti}(t) \times R_{GBTi}(t)))$$
(6)

where $R_{Ti}(t)$ and $R_{CBTi}(t)$ are the reliability of transformer i and of the main circuit breaker of transformer i, respectively, during the period t.

Second Group (SG)

The feeders are connected in a parallel manner, indicating that

$$R_{SG}(t) = 1 - \prod_{i=1}^{10} (1 - R_{Fi}(t))$$
(7)

where $R_{Fi}(t)$ is the reliability of feeder i during the period of t.

The group of transformers and the group of feeders are connected in a series. The reliability function of the system is

$$R_{SYS}(t) = R_{FG}(t) \times R_{SG}(t)$$

$$(8)$$

Based on the values presented in Table 4 and by using Eqs. 3, 6, 7, and 8, we can calculate the system reliability for the times imposed from $t = 1$ to $t = 30$, where t is expressed in days. The data are presented in Table 5.

Table 5: Estimated reliability system values of the electric power distribution station for 30 days.

T(day)	R_{SYS}	T(day)	R_{SYS}
1	0.996500665	2	0.989498957
3	0.979827045	4	0.968407728
5	0.95585446	6	0.942558222
7	0.928774489	8	0.914675671
9	0.900381613	10	0.885977738
11	0.871526224	12	0.857073115
13	0.842652974	14	0.828292031
15	0.814010334	16	0.799823272
17	0.785742656	18	0.771777517
19	0.75793469	20	0.744219264
21	0.730634929	22	0.717184252
23	0.703868901	24	0.690689829
25	0.67764743	26	0.664741662
27	0.651972165	28	0.639338348
29	0.626839464	30	0.614474679

doi:10.1371/journal.pone.0069716.t005

doi:10.1371/journal.pone.0069716.t005

Limitations of the Study, Open Questions, and Future Work

We believed that the limitations are:

- For the sake of brevity, we restrict our investigation to one electric power distribution station.
- We have used the data for five consecutive years.
- • We used EasyFit software for our investigation.
- The study focused in details inside the electric power distribution station, without return to the source. Note that if the source feeds the electric power distribution station by low energy (less than 33 kilovolt), this leads to a high temperature in the transformer which will cause a sudden stop of power station.

The present paper deals with the electric power distribution station as independent and separate components. If one take the data of failure rate (λ) for each components of this station and deal with λ by using the Markov model "Hidden Markov model." Then one may get better performance of the electric power distribution station. It is know that the Markov model dependent on the current state of the failure, rather returning to the history of the data [35].

The current paper may be extended to scrutinize preventive maintenance modelling and to estimate its effects on the components of the station. This might improve the supply of electrical energy and will reduce the operating cost of the power station.

CONCLUSION

The time between failures was analyzed to determine the best-fitting distribution. Using the distribution fitting software EasyFit, we determined that the most valid distribution is the Dagum distribution with a scale parameter $\beta = 90.001$, shape parameters $k = 0.33998$ and $\alpha = 2.4011$. The reliability value for the system on the first day was 0.99. If the station works for 30 days, the reliability value of station was decreased to 0.61. The value of the reliability function was declined by 38.2% in 30 days. This percentage indicates that the electric power distribution station studied in this paper exposed to fail close together in time, even in the same part. This leads to two possibilities; the first is that, the maintenance staff and engineers are not doing their work at best performance, the second is that parts for the electric power distribution station, consisting of 14 components are not good and are exposed to crash shortly after repairs. The first reason can be ignored, because, the field visits by researcher can emphasize the expertise of engineers and maintenance workers in the completion of repairs and maintenance in record time and dynamically good as mentioned in the records. The main problem is that the sum of component is not good. With reference to the case of calculating the Eqs. 6 and 7 for the reliability value for each component, it can be seen that the items CBT_1, CBF_2, CBF_3, CBF_4, CBF_5, CBF_6, CBF_7 and CBF_9 have the largest percentage in the value of reliability. Their reliability values after 30 days become as, 0.42, 0.45, 0.40, 0.37, 0.47, 0.41, 0.40, and 0.38 respectively. These values show that the components of the electric power distribution station continuously deteriorate because of aging. Furthermore, these components CBT_1, CBF_2, CBF_3, CBF_4, CBF_5, CBF_6, CBF_7 and CBF_9 had the highest critical value and must be changed with new items as soon as possible.

AUTHOR CONTRIBUTIONS

Conceived and designed the experiments: FMA. Performed the experiments: FMA. Analyzed the data: FMA GSH. Contributed reagents/materials/analysis tools: FMA GSH. Wrote the paper: FMA AB GSH. Data collected from the electricity company: FMA.

REFERENCES

1. Billinton R, Allan RN (1996) Reliability Evaluation of Power Systems. New York, USA: Plenum Press.

2. Alwan FM, Baharum A, Hasson ST (2012) Reliability and failure analysis for high power station based on operation time. In: International Conference on Statistics in Science, Business, and Engineering (ICSSBE). IEEE, 370–373.

3. Garvin DA (1996) Competing on the eight dimensions of quality. IEEE Engineering Management Review 24: 15–23.

4. Zio E (2007) An introduction to the basics of reliability and risk analysis, volume 13. Danvers, USA: World Scientific Publishing Company.

5. Kullstam PA (1981) Availability, MTBF and MTTR for repairable M out of N system. IEEE Transactions on Reliability 30: 393–394. doi: 10.1109/tr.1981.5221134

6. Schneeweiss W (1981) Computing failure frequency, MTBF & MTTR via mixed products of availabilities and unavailabilities. IEEE Transactions on Reliability R-30: 362–363. doi: 10.1109/tr.1981.5221115

7. Holtz J, Werner KH (1990) Multi-inverter ups system with redundant load sharing control. IEEE Transactions on Industrial Electronics 37: 506–513. doi: 10.1109/41.103455

8. Hamada MS, Wilson AG, Reese CS, Martz HF (2008) Bayesian Reliability. New York, USA: Springer.

9. Domma F, Giordano S, Zenga M (2011) Maximum likelihood estimation in dagum distribution with censored samples. Journal of Applied Statistics 38: 2971–2985. doi: 10.1080/02664763.2011.578613

10. Dagum C (2008) A new model of personal income distribution: specification and estimation. Modeling Income Distributions and Lorenz Curves 5: 3–25. doi: 10.1007/978-0-387-72796-7_1

11. Dagum C (1980) The generation and distribution of income, the lorenz curve and the gini ratio. Economie Applique 33: 327–367.

12. Meyer J, Ormiston MB (1983) The comparative statics of cumulative distribution function changes for the class of risk averse agents. Journal of Economic Theory 31: 153–169. doi: 10.1016/0022-0531(83)90026-1

13. Domma F, Giordano S, Zenga M (2009) The fisher information matrix in doubly censored data from the Dagum distribution. Technical report, Università Della Calabria.

14. Domma F (2002) Landamento della hazard function nel modello di dagum a tre parametri. Quaderni di Statistica 4: 1–12.

15. Frankel EG (1988) Systems Reliability and Risk Analysis. New York, USA: Springer.

16. Purcell IF, Poole-Wilson PA (1999) Heart failure: why and how to define it? European journal of heart failure 1: 7–10. doi: 10.1016/s1388-9842(98)00009-9

17. Carter ADS (1997) Mechanical Reliability and Design. New York: John Wiley & Sons, third edition.

18. Davison A, Kantor RN (1982) On the failure of readability formulas to define readable texts: A case study from adaptations. Reading Research Quarterly 17: 187–209.

19. Carter ADS (1986) Mechanical Reliability. London: Macmilan Education Ltd, second edition.

20. Audin G (2002) Reality check on five-nines. Business Communications Review 5: 22–27.

21. Martz H, Waller R (1982) Bayesian Reliability Analysis. New York, USA: John Wiley & Sons, Inc.

22. Dinesh Kumar U, Knezevic J, Crocker J (1999) Maintenance free operating periodan alternative measure to mtbf and failure rate for specifying reliability? Reliability Engineering & System Safety 64: 127–131. doi: 10.1016/s0951-8320(98)00048-9

23. Bryant CM, Schmee J (1979) Con_dence limits on mtbf for sequential test plans of mil-std 781. Technometrics 21: 33–42. doi: 10.1080/00401706.1979.10489720

24. Gaj F, Germani GU (2008) Dealing with availability in an international service management sce-nario, Springer. 235–244.

25. Dhillon BS (2006) Maintainability, Maintenance, and Reliability for Engineers. New York, USA: CRC Press.

26. Paul Barringer P (2004) Analyzer reliability. In: 18th International Forum for Process Analytical Chemistry. Virginia, USA, 2–9.

27. Alwan F, Baharum A, Hasson S (2012) The pdf fitting to time between failure for high power stations. Applied Mathematical Sciences 6: 6327–6339.

28. Goodman S (2008) A dirty dozen: Twelve p-value misconceptions. Seminars in Hematology 45: 135–140. doi: 10.1053/j.seminhematol.2008.04.003

29. Chaubey YP (1993) Resampling-based multiple testing: Examples and methods for p-value ad-justment. Technometrics 35: 450–451. doi: 10.1080/00401706.1993.10485360

30. Hedges SB (1992) The number of replications needed for accurate estimation of the bootstrap p value in phylogenetic studies. Molecular Biology and Evolution 9: 366–369.

31. Jha SK, Dwivedi A, Tiwari A (2011) Necessity of goodness of fit tests in research and development. International Journal of Computer Science and Technology 2: 135–141.

32. EASY-FIT (2013) Software for parameter estimation. University of London, UK. URL departments/math/kschittkowski/easyfit.htm.

33. Fakhraei Roudsari F (2011) Spatial, temporal and size distribution of freight train time delay in Sweden. Ph.D. thesis, KTH, Sweden.

34. Alwan FM, Baharum A, Hasson ST (2013) The performance of high-power station based on time between failures (TBF). Research Journal of Applied Sciences, Engineering and Technology 5: 3489–3498. doi: 10.1155/2013/583683

35. Ibe O (2008) Markov Processes for Stochastic Modeling. California, USA: Academic press.

Chapter 7

EXTRINSIC AND INTRINSIC FREQUENCY DISPERSION OF HIGH-K MATERIALS IN CAPACITANCE-VOLTAGE MEASUREMENTS

J. Tao [1], C.Z. Zhao[1,2,3], C. Zhao [2,3], P. Taechakumput [3], M. Werner [3,4], S. Taylor [3] and P. R. Chalker [4]

[1]Department of Microelectronics, Xi'an Jiaotong University, Xi'an 710016, China

[2]Department of Electrical and Electronic Engineering, Xi'an Jiaotong-Liverpool University, Suzhou 215123, China

[3]Department of Electrical Engineering and Electronics, University of Liverpool, Liverpool L69 3GJ, UK

[4]Department of Materials Science and Engineering, University of Liverpool, Liverpool L69 3GH, UK

ABSTRACT

In capacitance-voltage (C-V) measurements, frequency dispersion in high-k dielectrics is often observed. The frequency dependence of the dielectric constant (k-value), that is the intrinsic frequency dispersion, could not be assessed before suppressing the effects of extrinsic frequency dispersion, such as the effects of the lossy interfacial layer (between the high-k thin film and silicon substrate) and the parasitic effects. The effect of the lossy interfacial layer on frequency dispersion was investigated and modeled based on a dual frequency technique. The significance of parasitic effects (including series resistance and the back metal contact of the metal-oxide-semiconductor (MOS) capacitor) on frequency dispersion was also studied. The effect of surface roughness on frequency dispersion is also discussed. After taking extrinsic frequency dispersion into account, the relaxation behavior can be modeled using the Curie-von Schweidler (CS) law, the Kohlrausch-Williams-Watts (KWW) relationship and the Havriliak-Negami (HN) relationship. Dielectric relaxation mechanisms are also discussed.

INTRODUCTION

With increasing demand for higher speed and device density, the device dimensions in Si complementary-metal-oxide-semiconductor (CMOS) based integration circuits are continually being scaled down, following what is termed as Moore's law. The integrated circuit fabrication based on metal-oxide-semiconductor field-effect transistor (MOSFET) relies on thermally grown amorphous SiO_2 as a gate dielectric [1,2,3]. However, according to the International Technology Roadmap for Semiconductors (ITRS), CMOS technology could be extended to 14 nm nodes by 2020 by adopting novel device structure and new materials. The physical gate length and printed gate length of the device can be scaled down to 6 nm and 9 nm, respectively [4]. The rapid shrinking of feature size of transistors has forced the gate channel length and gate dielectric thickness on an aggressive scale. As the thickness of SiO_2 gate dielectric thin films used in metal-oxide-semiconductor (MOS) devices was reduced towards about 1 nm, the gate leakage current level became unacceptable. Below the physical thickness of 1.5 nm, the gate leakage current exceeds the specifications. To overcome this leakage problem, high-k materials were introduced because they allow the physical thickness of the gate stack to be increased but keep the equivalent oxide thickness (EOT) unchanged. Hence, the gate leakage was found to be reduced by two to three orders of magnitude.

On the other hand, capacitance-voltage (C-V) measurements are the fundamental characterization technique for MOS devices for the extraction of the oxide thickness [5], the maximal width of the depletion layer, interface trap densities [6], channel length [7], mobility [8], threshold voltage, bulk doping profile [9], and the distribution of the charges in dielectrics, which is used to evaluate the characterization of the interface states between the substrate and dielectric. Frequency dispersion in SiO_2 has frequently been observed in C-V measurements [10,11]. Several models and analytical formulae have been thoroughly investigated for correcting the data from measurement errors. Attention has been given to eliminate the effects of series resistance [12], oxide leakage, undesired thin lossy interfacial layer between oxide and semiconductor [13], surface roughness [14], polysilicon depletion [15,16,17] and quantum mechanical effect [18,19,20,21].

In this paper, the extrinsic and intrinsic causes of frequency dispersion during C-V or C-f (capacitance-frequency) measurements in high-k thin films were investigated. In order to reconstruct the measured C-V curves for any given measurement data, parasitic components including imperfection of the back contact and silicon series resistance which was one of the extrinsic causes of frequency dispersion must be taken into account. The corrected capacitance was provided following related models. Furthermore, another extrinsic cause of frequency dispersion, lossy interfacial layer effect, on high-k MOS capacitances was investigated for zirconium oxides and then a four-element circuit model was introduced. On the other side, frequency dispersion from the effect of surface roughness was best demonstrated in ultra-thin SiO_2 MOS devices [14] while the analysis of the $La_xZr_{1-x}O_{2-\delta}$ thin film and $Ce_xZr_{1-x}O_{2-\delta}$ thin film led to the conclusion that surface roughness was not responsible for the observed frequency dispersion for the thick high-k dielectric thin films. The polysilicon depletion effect and quantum confinement should be also considered. After taking into account all extrinsic causes of frequency dispersion mentioned above, the intrinsic effect (dielectric relaxation) of high-k dielectric thin films arose and several dielectric relaxation models were discussed. The dielectric relaxation results of $Ce_xZr_{1-x}O_{2-\delta}$, $LaAlO_3$, ZrO_2 and $La_xZr_{1-x}O_{2-\delta}$ thin films could be described by the Curie-von Schweidle (CS) law, the Kohlrausch-Williams-Watts (KWW) and the Havriliak-Negami (HN) relationship, respectively. The higher k-values were obtained from $La_xZr_{1-x}O_{2-\delta}$ and $Ce_xZr_{1-x}O_{2-\delta}$ thin films with the low lanthanide concentration levels (e.g., x ~ 0.1) where the more severe dielectric relaxation was observed. The causes of the dielectric relaxation were discussed in terms of this observation.

EXPERIMENTAL

The C-V and C-f measurements system consists of two Agilent precision LCR meters (4284A and 4275A), a desktop computer and a manual probe station. The MOS devices were wafer-probed on the probe station's loading platform and were connected from Agilent 4284A/4275A to the desktop computer and the probe station together through a GPIB interface, as shown in Figure 1. The data measured from the LCR meters were transferred back to the computer and saved to obtain the C-V curves automatically.

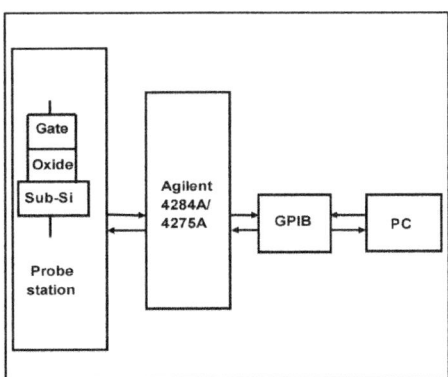

Figure 1: Capacitance-voltage (C-V) measurement system of metal-oxide-semiconductor (MOS) devices. A MOS device was located on the manual probe

station which was connected to the LCR meters (Agilent 4284A/4275A). The LCR meters were controlled by a desktop computer through a GPIB interface. The C-V measurement data extracted from the LCR meters were transferred back to the computer and saved to obtain the C-V curves automatically.

The structure of the MOS device shown in Figure 1 is similar to planar capacitors which are formed by metal and dielectric. The differential capacitance of a MOS capacitor is:

$$C = A\frac{dQ_G}{dV_G} = \frac{i_{ac}}{dV_{ac}/dt}$$

(1)

where Q_G and V_G are the charge area density and voltage on the metal electrodes, A is the metal electrode area, dV_{ac}/dt is the AC voltage change, and i_{ac} is the AC current. The capacitance of a MOS device was obtained by Agilent 4284A/4275A, which provided a small signal voltage variation rate (dV_{ac}/dt) and measured the small signal current (i_{ac}) flowing through the MOS device to calculate the differential capacitance of the MOS device according to Equation (1) [22,23]. For the Agilent 4284A/4275A precision LCR meters, there are two models used to calculate the device capacitance. One is the series model and the other is the parallel model, as shown in Figure 2. The parallel model was used in the following C-V and C-f measurements. In Figure 2, C_m is the measured capacitance. R_m and G_m are the measured resistance and conductance respectively. C_D is the depletion capacitance and Y_{it} is the admittance due to interface states of the MOS device, respectively. C_{ox} represents the actual frequency independent capacitance.

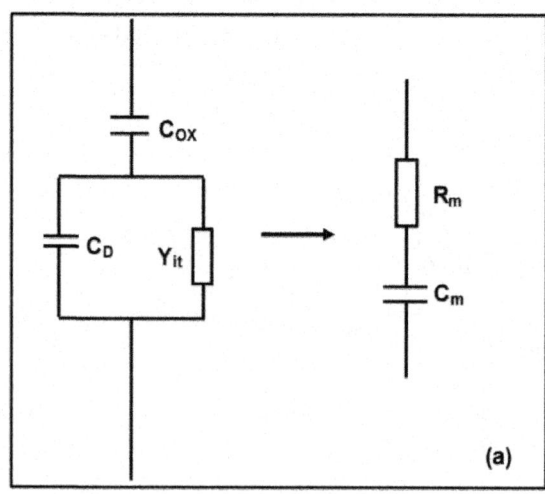

(a)

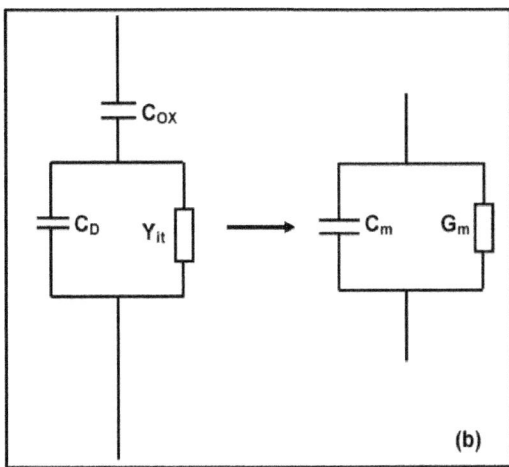

Figure 2: Conventional LCR meters typically measure the device capacitance based on (a) Series capacitance model or (b) Parallel capacitance model. C_m is the measured capacitance. R_m and G_m are the measured resistance and conductance respectively. C_D is the depletion capacitance and Y_{it} is the admittance due to interface states of the MOS device, respectively.

However, the influence of the leakage current of oxides to i_{ac} in the C-V and C-f measurements of MOS devices by the LCR meters should be taken into account. Especially crystalline thin films exhibit significantly higher leakage current than amorphous thin films, which could be due to the leakage pathway introduced from the grain boundaries and the local defects [24,25]. An approximation for the percentage instrumental error was given by the formula $0.1 \times \sqrt{1 + D^2}$, where D is a dissipation factor. If the instrumentation error is less than 0.3%, the leakage current in the MOS device is negligible [13]. In the following C-V and C-f measurements, the leakage current in high-k thin films was so small that it was not a contributing factor to frequency dispersion [26].

High-k dielectrics, $LaAlO_3$, ZrO_2, $Ce_xHf_{1-x}O_{2-x}$ and $La_xZr_{1-x}O_{2-\delta}$ thin films, were deposited on n-type Si (100) substrates using liquid injection atomic layer deposition (ALD), carried out on an Aixtron AIX 200FE AVD reactor fitted with the "Trijet" ℘ liquid injector system [27]. The doping level of $Ce_xHf_{1-x}O_{2-x}$ thin film and $La_xZr_{1-x}O_{2-\delta}$ thin film was varied up to a concentration level of 63%, i.e., x = 0.63. The interfacial layer between the high-k thin film and silicon substrate was a ~1 nm native SiO_2 determined by cross-section transmission electron microscopy (XTEM). A thermal SiO_2 sample was grown using dry oxidation at 1100 °C to provide a comparison with the high-k stacks. MOS capacitors were fabricated by thermal evaporation of Au gates through a shadow mask with an effective area of 4.9×10^{-4} cm². The backside contact of selected Si wafers was cleaned with a buffer HF solution and subsequently a 200 nm thickness of Al film was deposited on it by thermal evaporation. Some selected samples of $Ce_xHf_{1-x}O_{2-x}$ thin films and $La_xZr_{1-x}O_{2-\delta}$ thin films were annealed at 900 °C for 15 min in a N_2 ambient to crystallize the thin films before metallization. All the other samples were annealed in forming gas at 400 °C for 30 min. The C-V or C-f curves of $Ce_xHf_{1-x}O_{2-\delta}$, $La_xZr_{1-x}O_{2-\delta}$, ZrO_2, $LaAlO_3$ and thermal SiO_2 thin films were measured to investigate their electrical properties. X-ray diffraction (XRD), XTEM and atomic force microscopy (AFM) of $La_xZr_{1-x}O_{2-\delta}$ thin films and $Ce_xHf_{1-x}O_{2-\delta}$ thin films were used to investigate their physical properties.

RESULTS AND DISCUSSION

Frequency dispersion was categorized into two parts: extrinsic causes and intrinsic causes. Section 3.1 presented the extrinsic frequency dispersion. After analyzing the C-V curves of SiO_2 MOS capacitors (MOSC), the parasitic effect is introduced in Section 3.1.1. Dispersion could be avoided by depositing an Al thin film at the back of the silicon substrate. The correction models were able to minimize the dispersion as well. The existence of frequency dispersion in the $LaAlO_3$ sample is discussed in Section 3.1.2, which is mainly due to the effect of the lossy interfacial layer between the high-k thin film and silicon substrate on the MOSC. Relative thicker thickness of the high-k thin film than the interfacial layer significantly prevented frequency dispersion. Also, extracted C-V curves were reconstructed by mathematic correction models. Frequency dispersion from the effect of surface roughness was represented in an ultra-thin SiO_2 MOS device, which is discussed in Section 3.1.3. Furthermore, the surface property of the $La_xZr_{1-x}O_{2-\delta}$ thin films is studied. In Section 3.1.4 two further potential extrinsic causes: polysilicon depletion effect and quantum mechanical confinement, for frequency dispersion are considered. After careful considerations of extrinsic causes for frequency dispersion, intrinsic frequency dispersion is analyzed in Section 3.2. Section 3.2.1 describes the frequency dependence of k-value in $La_xZr_{1-x}O_2/SiO_2$ and $Ce_xHf_{1-x}O_{2-\delta}/SiO_2$ stacks. In order to interpret intrinsic frequency dispersion, several dielectric relaxation models are introduced in Section 3.2.2 for high-k materials with specified fitting parameters. Last but not least, three possible causes of the dielectric relaxation for the $La_xZr_{1-x}O_{2-\delta}$ dielectrics are proposed in Section 3.2.3. The effects of the cation segregation caused by annealing and rapped electrons on the dielectric relaxation were negligible. However, a decrease in crystal grain size may be responsible for the increase in the dielectric relaxation.

Extrinsic Causes of Frequency Dispersion During C-V Measurement

Several reasons for unwanted frequency dispersions in SiO_2 have been investigated, such as surface roughness [14], polysilicon depletion [15,16,17], quantum confinement (only for an ultra-thin oxide layer) [18,19,20,21], parasitic effect (including series resistance, back contact imperfection and cables connection) [28,29,30], oxide tunneling leakage current (direct tunneling current, F-N tunneling etc.) [31], unwanted interfacial lossy layer [13] and dielectric constant (k-value) dependence (dielectric relaxation) [26]. The extrinsic frequency dispersion is discussed firstly in Section 3.1. The extrinsic causes of frequency dispersion during C-V measurement in high-k thin film, which were investigated step by step before validating the effects of k-value dependence, were parasitic effect, surface roughness, and lossy interfacial layer. The other causes like tunneling leakage current and quantum confinement are negligible if the thickness of the high-k thin film is high enough. Polysilicon depletion effects were not considered due to the fact that metal gates were used here. The C-V results of high-k or SiO_2 based dielectrics are shown in Figure 3, Figure 4 and Figure 5, respectively. The parasitic effect (including back contact imperfection R_s', C_s', cables R_s'', C_s'' and substrate resistance R_s), the lossy interfacial layer effect C_i, G_i (between the high-k thin film and silicon substrate), polysilicon depletion effect and surface roughness on high-k thin films are summarized in detail in Figure 6.

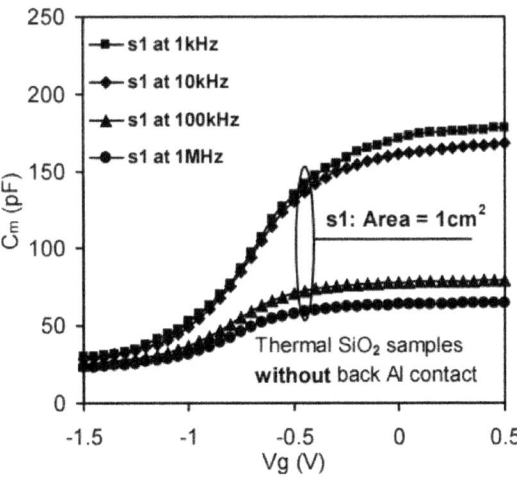

Figure 3: Frequency dispersion in C-V measurements observed in the thermal oxide (SiO_2) sample. In the absence of a substrate back Al contact, dispersion was evident in the sample with a small substrate area of $1 cm^2$ [32].

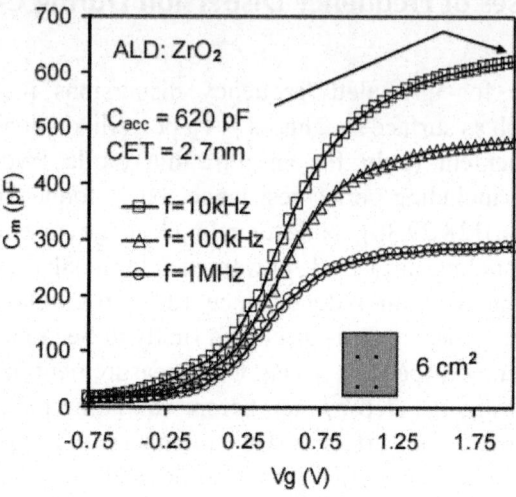

Figure 4: Presence of frequency dispersion in ZrO_2 samples at different frequencies (10kHz, 100kHz and 1MHz). The shadowed boxes indicate the presence of metal Al contact at the back of silicon substrates with an effective area of 6 cm² and the capacitance equivalent thickness (CET) is 2.7 nm. C_{acc} is the capacitance in the accumulation range [32].

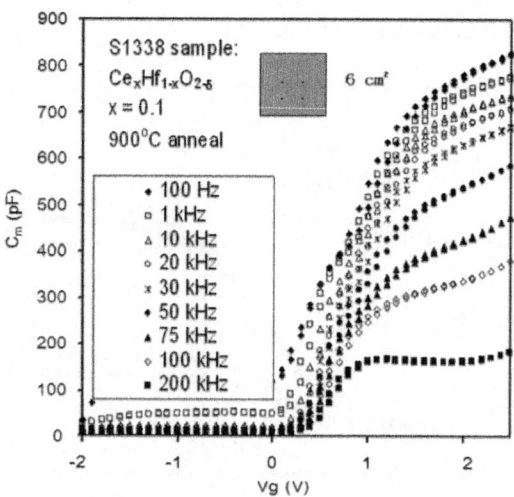

Figure 5: C-V curves from a $Ce_xHf_{1-x}O_{2-\delta}$ thin film at different frequencies (from 100 Hz to 200 kHz). Frequency dispersion could still be observed regardless of the interfacial layer effect of MOS structures and parasitic effects (caused by substrate resistance, back contact imperfection and cables). This kind of dispersion was caused by the frequency dependence of the k-value (dielectric relaxation) [33].

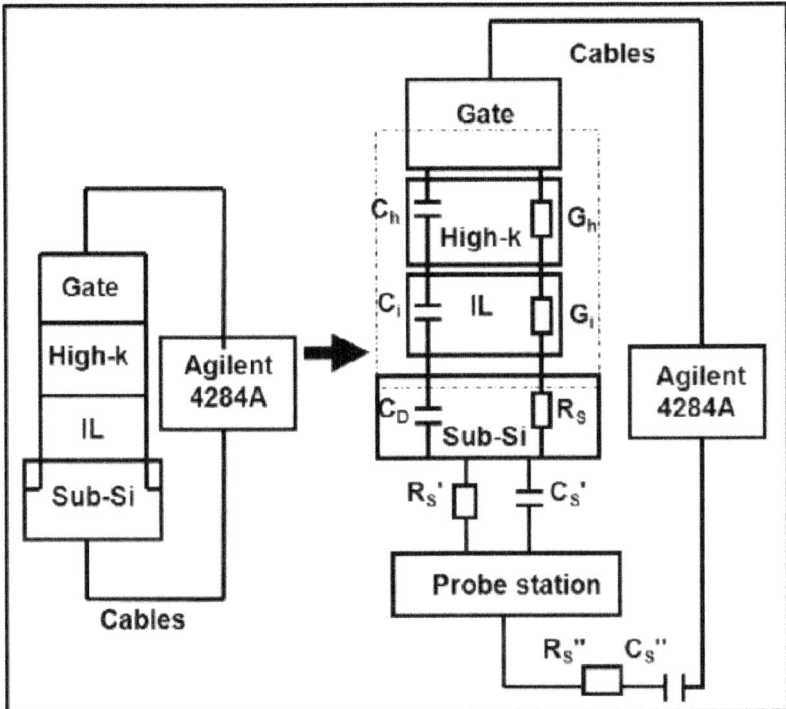

Figure 6: Causes of frequency dispersion during C-V measurement in the high-k thin film were the parasitic effect (including back contact imperfection resistance R_S' and capacitance C_S', cables resistance R_S'' and capacitance C_S'', substrate series resistance R_S and depletion layer capacitance of silicon C_D) and the lossy interfacial layer effect (interfacial layer capacitance C_i and conductance G_i). The dashed box includes surface roughness effect, polysilicon depletion effect, high-k capacitance C_h, high-k conductance G_h, the lossy interfacial layer capacitance C_i and conductance G_i. The oxide capacitance C_{ox} consists of the high-k capacitance C_h and the lossy interfacial layer capacitance C_i.

Parasitic Effect

Parasitic effects in MOS devices included parasitic resistances and capacitances such as bulk series resistances, series contact, cables and many other parasitic effects [34]. Five different sources of parasitic series resistance have been suggested [35]. However, only two of them which have practical importance are listed as follows: (1) the series resistance R_S of the quasi-neutral silicon bulk between the back contact and the depletion layer edge at the silicon surface underneath the gate; and (2) the imperfect contact of the back of the silicon wafer. Frequency dispersion caused by the parasitic effect is shown inFigure 3.

The significance of the series resistance effect, which was commonly due to silicon bulk resistance and back contact imperfection, was best demonstrated in thermal SiO_2 MOS capacitors, since in this case the effect of the lossy interfacial layer between the bulk dielectric and silicon substrate can be neglected. The thickness of thermal SiO_2 was thick enough to allow the tunneling leakage current to be neglected. [36,37]. Frequency dispersion in the SiO_2 capacitor was only observed in samples with small substrate effective areas as depicted in Figure 7a (closed symbols extracted from Figure 3). In addition, the measured results were also no longer reproducible for small samples in the absence of Al back contacts, as shown in Figure 7b (the closed symbols). It therefore impacted the measurement reliability.

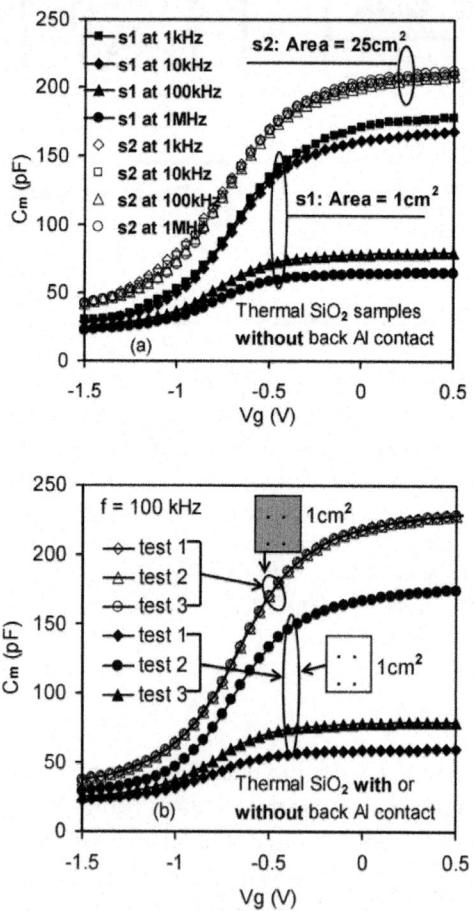

Figure 7: Frequency dispersion in C-V measurements observed in thermal oxide (SiO_2) samples. (a) In the absence of substrate back Al contact, dispersion was evident only in the sample with a smaller substrate area (denoted by s1); (b) The reproducibil-

ity of the tested devices in both the presence and absence of back metal contact. Both of the sample sets were measured three times within 24 hours. Closed symbols (e.g., ▲) signified the C-V results from the sample without back Al contact (indicated by a blank square), while the opened symbols (e.g., ○) showed the C-V results from the other sample with back Al contact (indicated by a shadow square) [32].

In order to reconstruct the measured C-V curves for any given measurement data in the frequency domain for SiO_2, one must take into account the parasitic components that may arise due to the silicon series resistance and the imperfection of the back contact. A correction may then be applied for the measured C-V curves in order to obtain their true values. Figure 8a shows an equivalent circuit of an actual case in comparison with the measurement mode, where C_{ox} represents the actual frequency independent capacitance across the SiO_2 gate dielectric, R_S includes both the bulk resistance in the silicon substrate and contributions from various contact resistances and cable resistances. The presence of the back contact capacitance and contributions from cable capacitance were also modeled by a capacitance. C_S, C_C, G_C, C_m, G_m refer to corrected (without the effect of the parasitic components R_S and C_S) measured capacitance and conductance, respectively. Following Kwa [13], the corrected capacitance C_C was given by [32]:

$$C_C = \frac{\left(\omega^2 C_m C_p - G_m^2 - \omega^2 C_m^2\right)\left(G_m^2 + \omega^2 C_m^2\right)C_p}{\omega^2 C_p^2 \left[G_m(1 - G_m R_S) - \omega^2 C_m^2 R_S\right]^2 + \left(\omega^2 C_m C_p - G_m^2 - \omega^2 C_m^2\right)^2} \qquad (2)$$

$$C_p = \frac{C_{ox}(G_{ma}^2 + \omega^2 C_{ma}^2)}{\omega^2(C_{ma}^2 C_{ox} - C_{ma}^2) - G_{ma}^2} \qquad (3)$$

$$R_S = \frac{G_{ma}}{G_{ma}^2 + \omega^2 C_{ma}^2} \qquad (4)$$

where C_{ma} and G_{ma} are the capacitance and conductance measured in strong accumulation. The measured capacitance can be recovered, independently of the measured frequencies, by applying the correction according to the model as depicted inFigure 8b. Alternatively, the parasitic effects can simply be minimized by depositing an Al thin film at the back of the silicon substrate (open symbols in Figure 7b and solid line in Figure 8b). In summary, it has been demonstrated that once the parasitic components are taken into account, it is possible to determine the true capacitance values free from errors. Therefore, the measurement system reliability can be maintained.

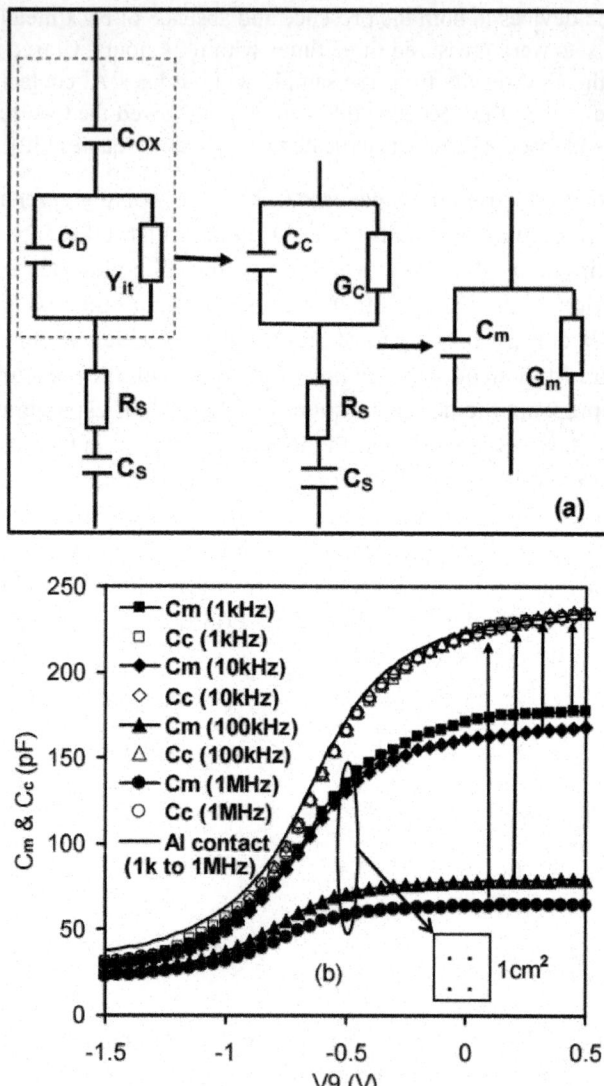

Figure 8: Effects of series resistance and back contact imperfection. (a) Equivalent circuit model, taking into account the presence of parasitic components from series resistance, cables and back contact imperfection (with the addition of the C_S and R_S). C_D is the depletion capacitance of silicon and Y_{it} is the admittance due to the interface states between SiO_2 and silicon substrate, respectively. C_{ox} is the oxide capacitance; (b) Extracted C_C-Vg curves based on measured data C_m and G_m using Equation (2). Dispersions disappear after considering C_S and R_S or depositing back Al contact (solid line). The blank square shows the tested device without back Al contact on silicon substrate. The effective substrate area is $1cm^2$ [32].

Lossy Interfacial Layer Effect

Concerning Figure 4, it should be noted that the dispersion was not caused by parasitic effects, since this sample had a large substrate area and an Al thin film was deposited on the back of the wafer. Subsequently, the effect of the lossy interfacial layer between the high-k thin film and silicon substrate on the high-k MOSC was investigated. The absence of frequency dispersion observed in Figure 9 may be explained in terms of the relative thickness of the high-k thin film compared to the interfacial layer. For the sample for Figure 9 the interfacial layer thickness (~1 nm) was negligible compared with the capacitance equivalent thickness (CET)¾!h of ~ 21 nm. Therefore in this case the high-k layer capacitance was much less than the interfacial layer capacitance (i.e., $C_h \ll C_i$) and the effect of C_i on C_m was eliminated. Furthermore the effect of the lossy interfacial layer conductance G_i on frequency dispersion can be suppressed by replacing the native SiO_2 by a denser SiO_2 thin film. In Figure 4, the frequency dispersion effect was significant even with the Al back contact and the bigger substrate area. In this case, C_h (CET = 2.7 nm) was comparable with C_i (~1 nm native SiO_2) and the frequency dispersion effect was attributed to losses in the interfacial layer capacitance, caused by interfacial dislocation and intrinsic differences in bonding coordination across the chemically abrupt ZrO_2/SiO_2 interface.

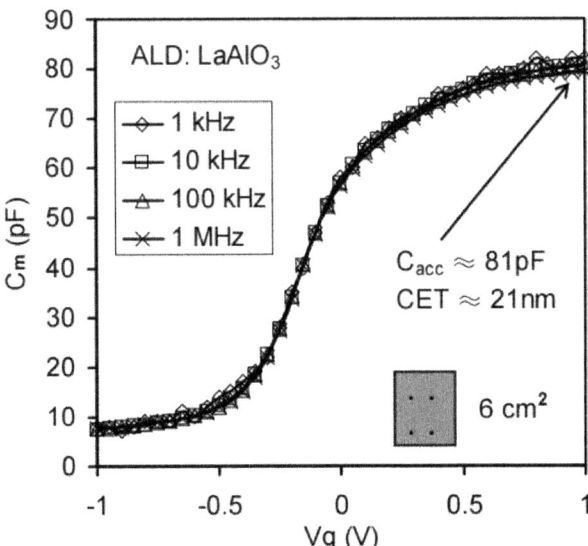

Figure 9: High frequency C-V results of $LaAlO_3$ thin film. The absence of frequency dispersion in the $LaAlO_3$ sample is observed with an effective area of 6 cm² with back Al contact. C_{acc} is the capacitance in the accumulation range [32].

Based on the above explanation, Figure 10a showed a four-element circuit model for high-k stacks, adapted from a dual frequency technique [10], with the capacitance value reconstructed from the loss. The expression for the corrected capacitance, C_c, was [32]:

$$C_C = \frac{\Delta^2 \left(\omega_1^2 - \omega_2^2 \right)}{\left[I_{m2}\omega_2\omega_1^2 \left(\Delta^2 + \omega_2^2 \right) - I_{m1}\omega_1\omega_2^2 \left(\Delta^2 + \omega_1^2 \right) \right]}$$

$$\Delta = \frac{\omega_1 I_{m1} - \omega_2 I_{m2}}{R_{m1} - R_{m2}}, \quad I_{mj} = \frac{\omega_j C_{mj}}{\left(G_{mj}^2 + \omega_j^2 C_{mj}^2 \right)}, \quad R_{mj} = \frac{G_{mj}}{\left(G_{mj}^2 + \omega_j^2 C_{mj}^2 \right)} \ and \ j = 1,2.$$

(5)

where C_m and G_m are the measured capacitance and conductance and ω is the measurement angular frequency. At an angular frequency ω_j ($j = 1$ or 2), the measured capacitance and conductance are C_{mj} and G_{mj} respectively. Since the expression of C_C with respect to ω_j, C_{mj} and G_{mj} is complicated, three abstract parameters, Δ, I_{mj}, and R_{mj} have been introduced to reduce the expression of C_C. Figure 10b shows the corrected C-V curves from Figure 4, extracted using Equation (5). All of the extracted C-V curves closely align with one another over the three different frequency pairs to reconstruct the true capacitance values. This indicates that the presence of a lossy interfacial layer is also responsible for the effect of frequency dispersion in high-k stacks.

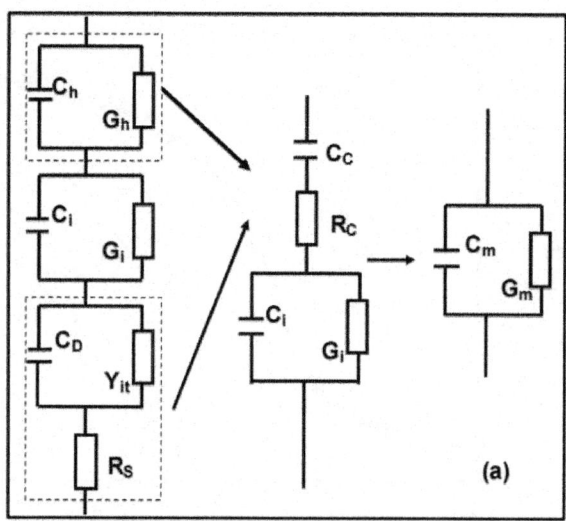

(a)

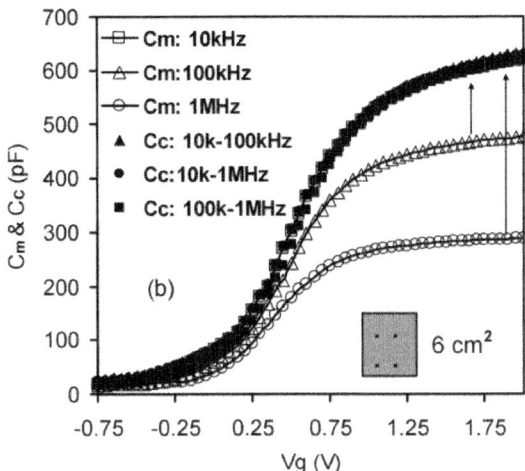

Figure 10: Effect of the lossy interfacial layer on high-k stacks. (a) Four-element equivalent circuit model for high-k stacks, taking into account the presence of the interfacial layer with the additional capacitance, C_I, and conductance, G_I, parallel circuit components. C_h and G_h represent the actual capacitance and conductance across the high-k dielectric. C_D is the depletion capacitance and Y_{it} is the admittance due to interface states, respectively; (b) Extracted C_C-Vg curves based on dual-frequency data from Figure 3 and the equivalent circuit model from Figure 10a [32].

Surface Roughness Effect

After taking the parasitic effects and the lossy interfacial layer effect into account, the unwanted frequency dispersion shown in Figure 5 may be caused by surface roughness. Frequency dispersion from the effect of surface roughness is best demonstrated in an ultra-thin SiO_2 MOS device [14]. In the following discussion, the effects of direct tunneling, series resistance and surface roughness on the capacitance were taken into account without considering quantum confinement and the polysilicon depletion effect. From Figure 11, the measured capacitance C_m is given by[38]:

$$C_m = \frac{C_{ideal}}{[(G+g)R_S + (\omega C_{ideal} R_S)]^2}$$

(6)

where G is the conductance due to a pure tunneling effect, g is the conductance due to the surface roughness effect, and R_S is the series resistance. From Equation (6), the real capacitance taking into account the surface roughness, C_{ideal}, can be calculated and it is free of frequency [38]. It was found that the surface roughness affects frequency dispersion when the thickness of ultra-thin oxides is ~1.3nm. To investigate whether the unwanted frequency dispersion of the high-k materials in Figure 5 is caused by the surface roughness or not, the surface properties of the $La_xZr_{1-x}O_{2-\delta}$ thin films was studied using AFM. The typical AFM micrographs of the $La_xZr_{1-x}O_{2-\delta}$ annealed thin films (x = 0.35 and x = 0.09) are shown inFigure 12.

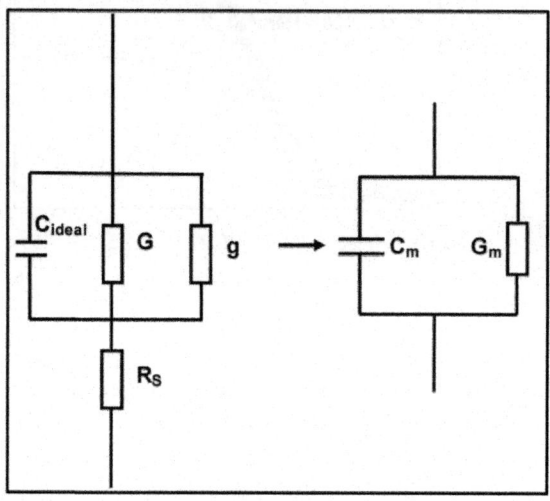

Figure 11: Equivalent circuit of the parallel mode of the measurement system. G is the conductance due to pure tunneling effect. g is the conductance due to the surface roughness effect. R_S is the series resistance. Figure is taken from Reference 36.

(a)

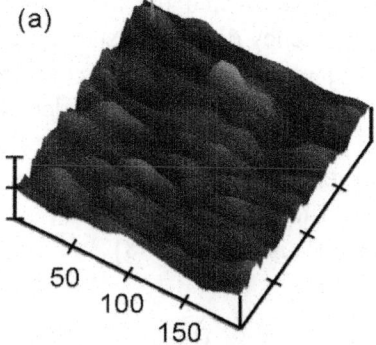

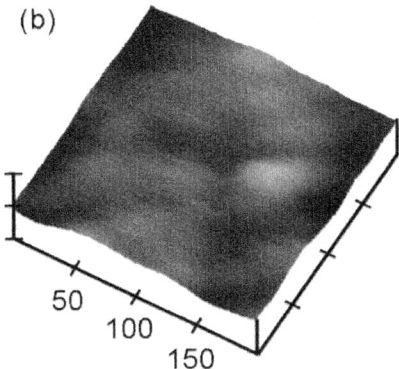

(b)

50
100
150

Figure 12: AFM micrographs of the surface of $La_xZr_{1-x}O_2$ annealed thin films. (a) x = 0.35; (b) x = 0.09 [33].

The Root Mean Square (RMS) roughness of the x = 0.35 thin film is 0.64 nm after annealing, as shown in Figure 12a. However no significant roughness was observed for the x = 0.09 thin film (RMS roughness of 0.3 nm), as shown in Figure 12b [26]. It means that the x = 0.35 thin film has more surface roughness than the x = 0.09 thin film. The applied frequency increased from 1 kHz to 1 MHz. The annealed thin film with x = 0.09 had large frequency dispersion where the capacitance decreased from 192 pF to 123 pF and the frequency changed from 1 kHz to 1 MHz. However, the annealed thin film with x = 0.35 showed small frequency dispersion where the capacitance decreased from 167 pF to 151 pF and the frequencies changed from 1kHz to 1MHz. Comparing these results from the C-V measurements in Figure 13, it leads to the conclusion that the surface roughness is not responsible for the observed frequency dispersion of the high-k dielectric thin films in Figure 13.

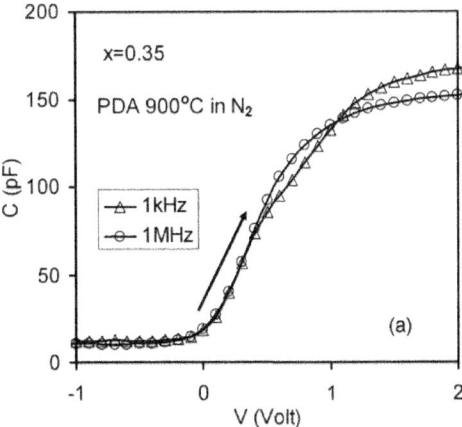

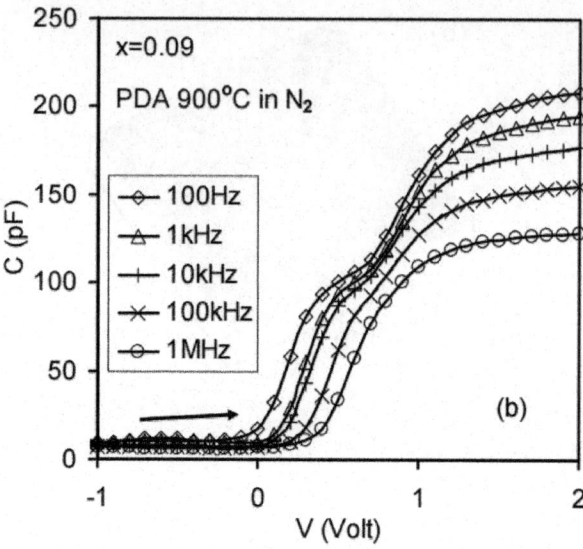

Figure 13: C-V results at different frequencies from the annealed $La_xZr_{1-x}O_{2-\delta}$ samples after back Al contact deposition and the effective substrate area was 6 cm^2: (a) x = 0.35; and (b) x = 0.09. Significant frequency dispersion was observed for the x = 0.09 annealed sample, but not for the x = 0.35 annealed sample [26].

Other Effects

There were two further potential causes of the observed frequency dispersion, polysilicon depletion effect and quantum mechanical confinement, for frequency dispersion which are not important for the samples in this paper. For thinner oxides, the poly depletion effect will become more significant leading to reduced surface potential, channel current, and gate capacitance. Furthermore, poly depletion will affect the extraction of the physical oxide thickness [39,40,41,42]. Some analysis and numerical results for the polysilicon depletion effect on the MOS device have been proposed [43,44]. The decrease in the gate capacitance caused by polysilicon depletion can be assumed as a cause of the increase in the effective gate oxide thickness. There are many surface potential models, which can be used to analyze the gate capacitance, solved by the Poisson Equation with boundary conditions to investigate the polysilicon depletion effect [45]. However, the polysilicon depletion effect was not under consideration for the samples used in this paper because the gates of the MOS capacitor samples were metal (Al or Au) fabricated by thermal evaporation through a shadow mask.

For oxide thicknesses down to 1~3 nm, the quantum mechanical effect should be taken into account [46,47,48]. There was a difference between the calculated capacitance and the measured capacitance with ultra-thin gate dielectrics. Quantum mechanical confinement would result in the continuous band being quantized into electric sub-band near the surface. The additional band bending confines the carriers to the narrow surface channel. The electron position changes and the peak of electron density is no longer in the silicon/silicon oxide interface, which would be further away from the surface in MOS devices [49,50]. However since the thickness of the high-k layer and interfacial layer is greater than 3 nm in the samples considered for this paper, the quantum mechanical effects were not considered.

Intrinsic Causes of Frequency Dispersion During C-V Measurements

Frequency Dependence of k-Value

Extrinsic causes of frequency dispersion during C-V measurements in high-k materials have been taken into account. Frequency dispersion can now solely be associated with the frequency dependence of the k-value in Figure 5, Figure 13 andFigure 14a. The frequency dependence of the k-value can be extracted as shown in Figure 14b, Figure 15 and Figure 16. The details are given below.

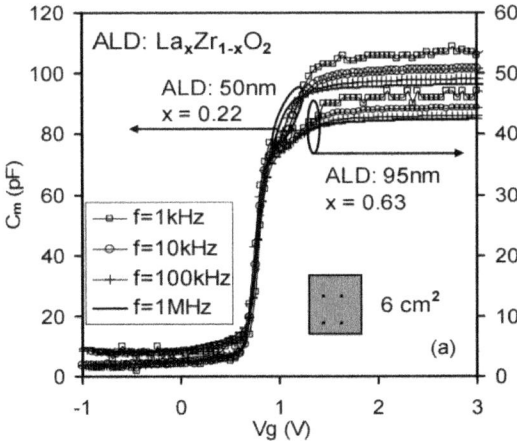

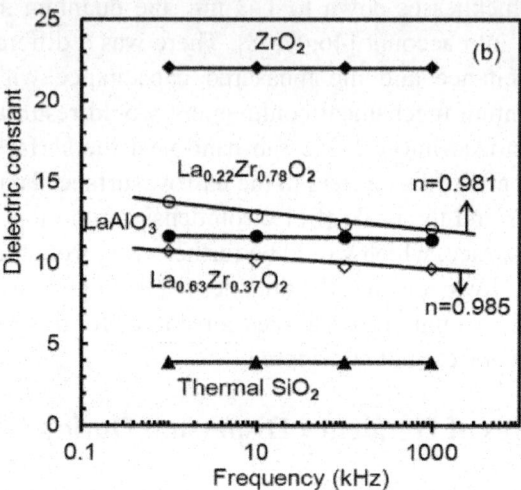

Figure 14: (a) Frequency dispersion in C-V measurements observed from $La_xZr_{1-x}O_2$ samples after back Al contact deposition and the effective substrate area was 6 cm². Therefore, all the extrinsic causes of frequency dispersion were excluded; (b) A summary of frequency dependence of k-value extracted from Figure 14a, Figure 7(SiO₂), Figure 9 (LaAlO₃), and Figure 10 (ZrO₂). No frequency dependence of k-value was observed for the LaAlO₃/SiO₂ and ZrO₂/SiO₂ stacks. The frequency dependence of the k-value was observed for the $La_xZr_{1-x}O_2$/SiO₂ stacks [32].

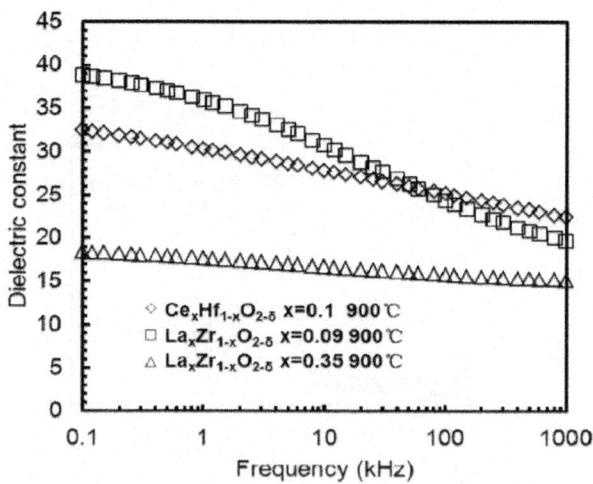

Figure 15: Frequency dependence of the k-value was extracted from C-f measurements of $La_{0.35}Zr_{0.65}O_{2-\delta}$ and $La_{0.09}Zr_{0.91}O_{2-\delta}$ thin films annealed at 900 °C, or extracted from Figure 13 (a,b). Frequency dependence of the $Ce_xHf_{1-x}O_{2-\delta}$ thin film was extracted from Figure 5 [33].

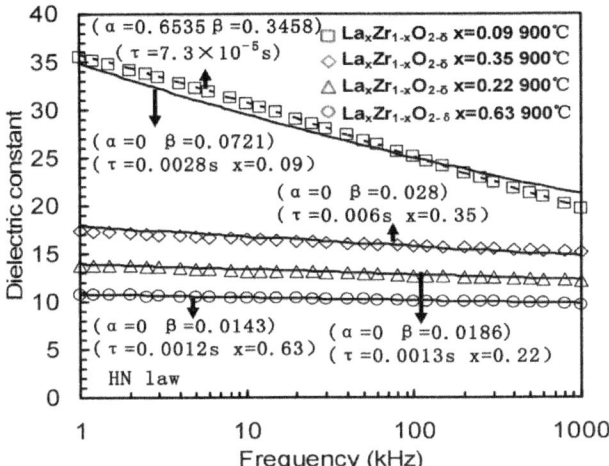

Figure 16: Frequency dependence of the k-value was extracted from C-f measurements observed in four $La_xZr_{1-x}O_{2-\delta}$ thin films. The square-symbols were measured from the $La_{0.09}Zr_{0.91}O_{2-\delta}$ sample. The diamond-symbols were measured from the $La_{0.35}Zr_{0.65}O_{2-\delta}$ sample. The triangle-symbols were measured from the $La_{0.22}Zr_{0.78}O_{2-\delta}$ sample. The circle-symbols were measured from the $La_{0.63}Zr_{0.27}O_{2-\delta}$ sample. Solid lines are from fitting results from the Cole-Davidson equation, while the dashed line is from the HN equation. The parameters α, β and τ are parameters from the Cole-Davidson or HN equation [32,33].

C-V data from the annealed thin films ($La_{0.22}Zr_{0.78}O_2$ and $La_{0.63}Zr_{0.37}O_2$) are given in Figure 14a. Figure 14b showed no frequency dependence of the k-value in $LaAlO_3/SiO_2$ and ZrO_2/SiO_2 stacks. However, the frequency dependence of the k-value was observed in $La_xZr_{1-x}O_2/SiO_2$ stacks. The k-values of $La_{0.22}Zr_{0.78}O_2$ and $La_{0.63}Zr_{0.37}O_2$ were observed and separately decreased from 13.5, 10.5 to 12 and 9.5 as the frequency increased from 1 kHz to 1 MHz. A constant frequency response was observed in thermal SiO_2, as shown in Figure 14b.

The k-f (k-value-frequency) data of the $Ce_xHf_{1-x}O_{2-\delta}$, $La_{0.35}Zr_{0.65}O_{2-\delta}$ and $La_{0.09}Zr_{0.91}O_{2-\delta}$ thin films are given in Figure 15. The zirconia thin film with a lanthanum (La) concentration of x = 0.35 showed that a k-value slowly decreased from 18 to 15 as the frequency increased from 100 Hz to 1 MHz. In contrast the lightly doped 9% sample had a sharp decreased k-value and suffered from a severe dielectric relaxation. A k-value of 39 was obtained at 100Hz, but this value was reduced to a k-value of 19 at 1 MHz. The 10% Ce doped hafnium thin film also had a k-value change from 33 at 100 Hz to 21 at 1 MHz. Figure 16 summarizes the frequency dependence of k-value of four $La_xZr_{1-x}O_2$ thin films from Figure 14 and Figure 15.

Many dielectric relaxation models were proposed to interpret intrinsic frequency dispersion which is also termed as frequency dependence of k-value. The fitted parameters of the dielectric relaxation models for Figure 16 are shown in Table 1. The details of the models are discussed below.

Table 1: The fitted parameters of the dielectric relaxation models for Figure 16

Models	Cole-Cole		Cole-Davidson		Havriliak-Negami		
Parameters	α	τ (s)	β	τ (s)	α	β	τ (s)
$La_xZr_{1-x}O_{2-\delta}$ $x=0.09$	0.75	3.9×10^{-7}	0.0721	0.0028	0.6535	0.3458	7.3×10^{-5}
$La_xZr_{1-x}O_{2-\delta}$ $x=0.35$	0.866	4.6×10^{-11}	0.028	0.006	0	0.028	0.006
$La_xZr_{1-x}O_{2-\delta}$ $x=0.22$	0.815	3.8×10^{-11}	0.0186	0.0013	0	0.0186	0.0013
$La_xZr_{1-x}O_{2-\delta}$ $x=0.63$	0.82	5.2×10^{-12}	0.0143	0.0012	0	0.0143	0.0012

Dielectric Relaxation Models and Data Fitting

In 1929, Debye [51] described a model for the response of electric dipoles in an alternating electric field. This model led to a description for the complex dielectric constant ε^*. The Debye equation and its real part are [51]:

$$\varepsilon^*(\omega) = \varepsilon_\infty + (\varepsilon_s - \varepsilon_\infty)/[1 + (i\omega\tau)] \tag{7}$$

$$\varepsilon'(\omega) = \varepsilon_\infty + (\varepsilon_s - \varepsilon_\infty)/[1 + \omega^2\tau^2] \tag{8}$$

where τ is called the relaxation time which is a function of temperature and it is independent of the time, angular frequency $\omega = 2\pi f$. At static conditions the dielectric behavior is characterized by the relative static dielectric permittivity ε_s, which is usually denoted as "static dielectric constant". ε_s is also defined as the zero-frequency limit of the real part, ε', of the complex permittivity. ε_∞ is the dielectric constant at ultra-high frequency. ε' is the k-value.

The Debye theory assumed that the molecules are spherical in shape and dipoles are independent in their response to the alternating field with only one relaxation time. The Debye equation (8) predicates ε' sharp decreases with frequency over a relatively small band width. Generally, the Debye theory of dielectric relaxation is utilized for particular type polar gases, dilute solutions of polar liquids and polar solids. [52] However, the dipoles for a majority of materials are more likely to be interactive and dependent in their response to the alternating field. Therefore, very few materials completely agree with the

Debye equation which has only one relaxation time. The Debye expression cannot interpret the data of polar dielectrics with a distribution of relaxation times (comparing to one relaxation time) [53]. For example, Figure 15 shows that the intrinsic frequency dispersion of the high-k materials ($La_xZr_{1-x}O_2$ and $Ce_xHf_{1-x}O_{2-\delta}$) occurred over a wide frequency range. The data was unable to be fitted with the Debye equation because the high-k materials have more than one relaxation time.

Since the Debye expression cannot properly predict the behavior of some liquids and solids such as chlorinated diphenyl at −25 °C and cyclohexanone at −70 °C [52], Cole K.S. and Cole R.H. proposed an improved Debye equation, known as the Cole-Cole equation, to interpret data observed on various dielectrics. Among relaxation frequencies Cole-Cole relaxation showed that ε' decreased more slowly with frequency than the Debye relaxation. By observing a large number of materials, they found that when the imaginary part (ε») was plotted versus ε', a curved arc resulted, whereas a semicircle was predicted by the Debye relation. The Cole-Cole equation can be represented by ε*(ω) [52]:

$$\varepsilon * (\omega) = \varepsilon_\infty + (\varepsilon_s - \varepsilon_\infty)/[1 + (i\omega\tau)^{1-\alpha}]$$

(9)

where τ is relaxation time and α is a constant for a given material, having a value $0 \leq \alpha \leq 1$. α = 0 for Debye relaxation. The real part of the Cole-Cole equation is:

$$\varepsilon'(\omega) = \varepsilon_\infty + (\varepsilon_s - \varepsilon_\infty)\frac{1 + (\omega\tau)^{1-\alpha}\sin\frac{1}{2}\alpha\pi}{1 + 2(\omega\tau)^{1-\alpha}\sin\frac{1}{2}\alpha\pi + (\omega\tau)^{2(1-\alpha)}}$$

(10)

The larger the value of α, the larger is the distribution of relaxation times. The Cole-Cole equation can be used to fit the dielectric relaxation results shown in Figure 16 of the $La_{0.91}Zr_{0.09}O_2$, $La_{0.22}Zr_{0.78}O_2$, $La_{0.35}Zr_{0.65}O_2$ and $La_{0.63}Zr_{0.37}O_2$ thin films and the fitted parameters are shown in Table 1. All of the data perfectly fitted, but the relaxation time was too small (e.g., 10^{-11}s), as shown in Table 1.

Davidson et al. [53] proposed the following expression (Cole-Davidson equation) to interpret data observed on propylene glycol and glycerol based on the Debye expression:

$$\varepsilon * (\omega) = \varepsilon_\infty + (\varepsilon_s - \varepsilon_\infty)/(1 + i\omega\tau)^\beta$$

(11)

where τ is the relaxation time and β is a constant for a given material. $0 \leq \beta \leq 1$ which controlled the width of the distribution and $\beta = 1$ for Debye relaxation. The smaller the value of β then the larger is the distribution of relaxation times [54]. For angular frequencies $\omega > 1/\tau$, the Cole-Davidson model exhibits an asymmetric broadening of the spectrum towards high frequency. The data of propylene glycol and glycerol can be fitted with the Debye formula in a low frequency region. However, at high frequencies, the Debye formula is no longer suitable for fitting. The data can be properly fitted by the Cole-Davidson formula instead. [55] This was reported as a limiting case to the Debye equation. The asymmetric loss factorε" was more seriously in error as the parameter β increased.

The real part of Equation (11) is given by [56]:

$$\varepsilon'(\omega) = \varepsilon_\infty + (\varepsilon_s - \varepsilon_\infty)(\cos\varphi)^\beta \cos\beta\varphi$$

(12)

$$\varphi = arctg(\omega\tau)$$

(13)

The Cole-Davidson equation could also fit the dielectric relaxation results shown in Figure 16 and the fitted parameters are shown in Table 1. However, the fitting for the $La_{0.91}Zr_{0.09}O_2$ thin films was not acceptable.

Both the Cole-Cole and Cole-Davidson equations are empirical and could be considered to be the consequence of the existence of a distribution of relaxation times rather than that of the single relaxation time (Debye equation). The physical reason for the distribution of relaxation times in the Cole-Cole and Cole-Davidson empirical equations is not yet clear. The reason for a distribution of relaxation times has been made in certain particular cases, e.g., the occurrence of protonic resonance (reported by Kliem and Arlt [57]) and the porosity effect (proposed by Cabeza et al. [58]).

In 1966, S. Havriliak and S. J. Negami reported the Havriliak-Negami (HN) equation which combined Cole-Cole and Cole-Davidson equations for twenty one polymers [59,60]. The HN equation is [60]:

$$\varepsilon^*(\omega) = \varepsilon_\infty + (\varepsilon_s - \varepsilon_\infty)/(1 + (i\omega\tau)^{1-\alpha})^\beta$$

(14)

The real part of the HN equation is given by [61]:

$$\varepsilon'(\omega) = \varepsilon_\infty + (\varepsilon_s - \varepsilon_\infty)\frac{\cos(\beta\phi)}{(1 + 2(\omega\tau)^{1-\alpha}\sin(\pi\alpha/2) + (\omega\tau)^{2(1-\alpha)})^{\beta/2}}$$

(15)

$$\phi = arctg \frac{(\omega\tau)^{1-\alpha} \cos\frac{1}{2}\pi\alpha}{1+(\omega\tau)^{1-\alpha} \sin\frac{1}{2}\pi\alpha}$$

(16)

where α and β are the two adjustable fitting parameters. α is related to the width of the loss peak and β controls the asymmetry of the loss peak [62]. In this model, parameters α and β could both vary between 0 and 1. The Debye dielectric relaxation model with a single relaxation time from Equation (14) $\alpha = 0$ and $\beta = 1$, the Cole-Cole model with symmetric distribution of relaxation times follows for $\beta = 1$ and $0 \leq \alpha \leq 1$, and the Cole-Davidson model with an asymmetric distribution of relaxation times follows for $\alpha = 0$ and $0 \leq \beta \leq 1$. The HN equation had two distribution parameters α and β but Cole-Cole and Cole-Davidson equations had only one.

This relaxation function had two intriguing features associated with it. First, and most importantly, it represented the experimental quantities almost within their reliability. Secondly, this function could be considered as a generalized way of writing the two known and well documented dispersion functions of Cole [52]. Hartmann et al. [62] have shown that the five parameters HN model used in the frequency domain can accurately describe the dynamic mechanical behavior of polymers, including the height, width, position, and shape of the loss peak.

The HN equation can be fitted for the dielectric relaxation results of the four $La_{0.91}Zr_{0.09}O_2$, $La_{0.22}Zr_{0.78}O_2$, $La_{0.35}Zr_{0.65}O_2$ and $La_{0.63}Zr_{0.37}O_2$ thin films more accurately than the Cole-Cole and Cole-Davidson equations which have only one distribution parameter. The fitting curves are shown in Figure 16. The fitting parameters of the $La_xZr_{1-x}O_2$ (x = 0.09, 0.22, 0.35 and 0.63) dielectrics are provided in Table 1.

From Table 1 and Figure 16, the fitting results of the Cole-Davidson equation showed that the asymmetry of the dielectric loss peak, β, increases with decreasing concentration, x, of La. To best fit the x = 0.09 sample, the width change of the loss peak α should be taken into account and, therefore, the HN Equation (15) should be used, where $\alpha = 0.6535$, $\beta = 0.3458$ and $\tau = 7.3 \times 10^{-5}$ s.

For the fitting of data in the time domain, an empirical expression was proposed by Kohlrausch, Williams and Watts, which is a stretched exponential function, $exp[-(t/\tau K)\beta K]$, [63] to be referred to later as the Kohlrausch-Williams-Watts (KWW) function. The equivalent of the dielectric response function is:

$$f(t) = d\Phi/dt \tag{17}$$

$$\Phi(t) = \exp[-(t/\tau_K)^{\beta_K}] \tag{18}$$

where the τ_K is the characteristic relaxation time, β_K is a stretching parameter, whose magnitude could vary from 0 to 1. For $\beta_K = 1$ the Debye process is obtained. In order to analyze the KWW law in the frequency domain, a Fourier transform is needed. The KWW function in the frequency domain is [63]:

$$\varepsilon^*(\omega) = \varepsilon_\infty + (\varepsilon_s - \varepsilon_\infty)\int_0^\infty \beta\tau^{-\beta_K}t^{\beta_K-1}\exp\left[-(t/\tau)^{\beta_K} - i\omega t\right]dt \tag{19}$$

The KWW law has been widely used to describe the relaxation behavior of glass-forming liquids and other complex systems [64]. The KWW law is not simply an empirical expression, but has a profound theoretical significance. Ngai et al.[65,66,67,68] developed a coupling model and derived the Kohlrausch function theoretically. It has already been pointed out by Yoshihara and Work [69] from their careful dielectric measurements that the HN equation can describe the complex permittivity of poly more precisely than the KWW function because the HN equation has two distribution parameters α and β but the KWW function has only one parameter β_K [70]. However, a possible relationship between α, β and β_K was hinted at by the results in Reference [71,72,73], where the following analytical relations could be derived:

$$\beta_K = [(1-\alpha)\beta]^{1/1.23} \tag{20}$$

For characteristic relaxation times, the relationship of τ (the relaxation time of the HN equation) and τ_k is

$$\ln(\tau/\tau_K) = 2.6*(1-\beta_K)^{0.5}\exp(-3\beta_K) \tag{21}$$

where α and β are the distribution parameters of the HN equation and β_K is the distribution parameter of the KWW equation. For the shape parameters, there is a direct transformation from the HN parameters into the KWW parameter. It is well known that a Fourier transform is needed to analyze the KWW law in the frequency domain. However, there is no analytic expression for the Fourier transform of the KWW function in the frequency domain. Any Fourier transform of the KWW function in the frequency domain can be approximated by a HN function which has a more complex relaxation form, but not vice versa [74].

In time domain, the general type of dielectric relaxation can be also described by the Curie-von Schweidler (CS) law (the t^n behavior, $0 \le n \le 1$) [75,76]. After a Fourier transform, the complex susceptibility CS relation is:

$$\chi_{CS} = A(i\omega)^{n-1}$$
(22)

where A and n are the relaxation parameters, ε_∞ is the high frequency limit of the permittivity, $\chi_{CS} = [\varepsilon_{CS} \times (\omega) - \varepsilon_\infty]/(\varepsilon_s - \varepsilon_\infty)$ is the dielectric susceptibility related to the CS law [53]. The value of the exponent (n) indicates the degree of dielectric relaxation [63,77]. The values obtained for the exponent n, showed that a weak dependence of the permittivity on frequency was observed [78]. A n−1 value of zero would indicate that the dielectric permittivity is frequency independent (no dielectric relaxation) [79].

The CS behavior is shown to be faster than the HN function at short times and slower than the HN function at long times. Although the CS relation is empirical, there are many models which link it to physical properties. The majority of these models are based on the presence of compositional or structural inhomogeneities and/or many body effects [80].

To best fit the experimental data, the frequency dependence of complex permittivity $\varepsilon^*(\omega)$ can be combined with the CS law and the KWW law [65]:

$$\varepsilon^*(\omega) = \varepsilon_\infty + \chi_{CS}(\omega) + \chi_{KWW}(\omega) - i\sigma/(\omega\varepsilon_s)$$
(23)

where ε_∞ is the high frequency limit permittivity, ε_s is the permittivity of free space, σ is the dc conductivity, $\chi_{KWW} = [\varepsilon_{KWW}^*(\omega) - \varepsilon_\infty]/(\varepsilon_s - \varepsilon_\infty)$ is the dielectric susceptibility related to the KWW law [53].

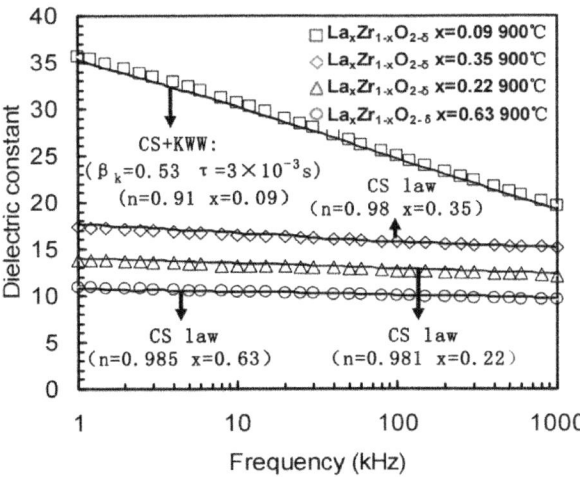

Figure 17: The measured data are the same as Figure 16. All solid lines are fitting results from the CS law or the combined CS+KWW laws. β_K, τ_K and n are parameters of the KWW law and CS law [26,32].

The dielectric relaxation data in Figure 16 were modeled by the CS law or the combined CS+KWW laws, as shown inFigure 17. The k-values of the LaxZr1−xO2 (x = 0.22, 0.35 and 0.63) dielectrics clearly show a power-law dependence on frequency known as the CS law, $k \propto f n-1$, $(0 \leq n \leq 1)$ [33,34]. For $La_x Zr_{1-x} O_{2-\delta}$ thin films with x = 0.63, x = 0.35 and x = 0.22 La content, the dielectric relaxation response could be fitted by the pure CS law, the n values were 0.981, 0.98 and 0.985 when the composition of La, x, were 0.22, 0.35 and 0.63, respectively.

However, for the x = 0.09 La content, the dielectric relaxation response could not be modeled by the pure CS law or the pure KWW law, but could be modeled by the combined CS+KWW law (Equation (23)). The relaxation parameter β_k and nwere 0.53 and 0.91, respectively, and the relaxation time τ_K was 3×10^{-3} s, as shown in Figure 17. From Figure 17, the exponent value n decreased withincreasing k-values.

Compared with Figure 16, it was found that the combined CS+KWW relaxation process (β_k = 0.53 and n = 0.91) could be substituted by the HN function where α and β were 0.6535 and 0.3458, respectively, in Figure 16 because both the HN function and the combined CS+KWW relationship both have two distribution parameters.

Dielectric Relaxation Mechanisms

XRD data from several doped thin films are given in Figure 18. All as-deposited samples were amorphous, but became tetragonal or cubic phase after annealing at 900 °C for 15 min. It was clear that a La concentration of x = 0.09 stabilized a mixed phase of zirconia of either the tetragonal or cubic phase, with some diffraction features from the monoclinic phase. For the hafnia thin films, Cerium (Ce) doping at a concentration of 10% stabilized the tetragonal or the cubic phase, and no monoclinic features were observed.

Doping hafnia and zirconia thin films with rare earth elements can stabilize the tetragonal or the cubic phase following annealing which enhances the k-value [26,34]. The highest k-value was obtained with lightly doped thin films, with a doping level of around 10% for both materials, as show in Figure 15 and Figure 16. The level of enhancement was closely related to the doping level. This experimental finding was in close agreement with the predictions of theoretical studies [81,82]. For low levels of doping (~10%): (1) k-values of 39 and 33 were obtained from $La_{0.09} Zr_{0.91} O_{2-\delta}$ thin film and $Ce_{0.1} Hf_{0.9} O_{2-\delta}$ thin film, respectively, as shown in Figure 15; (2) the tetragonal or cubic phase was formed, but was accompanied by significant dielectric relaxation. For high levels of doping (such as 35%), (1) no significant enhancement of the k-value was achieved; (2) the tetragonal or cubic phase was formed and no significant dielectric relaxation was observed, as shown in Figure 16 [33].

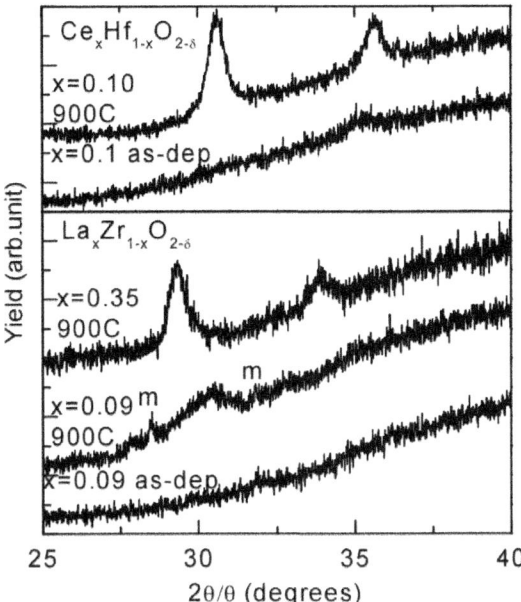

Figure 18: XRD data from La$_x$Zr$_{1-x}$O$_{2-\delta}$ thin films (bottom) and Ce$_x$Hf$_{1-x}$O$_{2-\delta}$ thin films (top), as-deposited and following annealing at 900 °C. As-deposited thin films were amorphous. In the annealed thin films, diffraction peaks were from the tetragonal or cubic phase. Data from the monoclinic phase was labeled, m [33].

Three possible causes of the dielectric relaxation for the La$_x$Zr$_{1-x}$O$_{2-\delta}$ dielectric are possible: (1) ion movement of unbounded La$^+$ or Zr$^+$ ions in the metal-oxide lattice resulting in dielectric relaxation [83]; (2) the combination of unbound metal ions with electron traps, generating dipole moments and inducing dielectric relaxation [84]; (3) a decrease in crystal grain size, causing an increase in the dielectric relaxation due to increased stresses [85,86]. It has been shown that the effect of the cation segregation caused by annealing and rapped electrons on the dielectric relaxation were negligible [26]. However, it has been reported that a decrease in crystal grain size can cause an increase in the dielectric relaxation in ferroelectric relaxor ceramics and this relaxation effect has been attributed to higher stresses in the smaller grains [85,86]. After annealing, the doping level affected the phase of the thin film crystallization and the size of the crystal grains formed that cause the dielectric relaxation. For a La concentration of x = 0.35 dielectric thin films with the 900 °C N$_2$-annealed containing ~15 nm crystals did not suffer from severe dielectric relaxation and a similar effect appeared to occur with the 900 °C air annealed, producing ~4 nm diameter equiaxed nanocrystallites within the thin film, which suffered from severe dielectric relaxation [87]. So, the cause of the dielectric relaxation is believed to be related to the size of the crystal grains formed during annealing and doping affects the size of the crystal grains formed.

CONCLUSIONS

In summary, in this paper, extrinsic and intrinsic frequency dispersion have been discussed in detail. Two causes of extrinsic frequency dispersion were investigated including the parasitic effect (series resistance, back contact imperfection, cables and connections) and the lossy interfacial layer. These effects were analyzed and modeled based on correction models. Secondly, the surface roughness was observed in ultra-thin dioxide thin films. However, aftsssser AFM micrographs were analyzed, the surface roughness was found not to be responsible for the observed frequency dispersion of the thick high-k dielectric thin films (>3 nm).

Lastly but not least, after causes for the extrinsic frequency dispersion were considered and determined, intrinsic frequency dispersion (dielectric relaxation) was found to be strongly related to the frequency dependence of the k-value on the high-k MOS capacitors. For low levels of doping (~10%), k-values of 39 and 33 were obtained from the $La_{0.09}Zr_{0.91}O_{2-\delta}$ thin film and $Ce_{0.1}Hf_{0.9}O_{2-\delta}$ thin film, respectively, at 100 Hz; while no significant enhancement of the k-value was achieved with high levels of doping (such as 35%).

The dielectric relaxation models in the frequency domain (such as the Cole-Cole equation, the Cole-Davidson equation, the HN equation) and in the time domain (such as the KWW law and the CS law) were comprehensively considered. The dielectric relaxation results of the $Ce_xZr_{1-x}O_{2-\delta}$, $LaAlO_3$, ZrO_2 and $La_xZr_{1-x}O_{2-\delta}$ thin films may be described by either the combined CS+KWW laws or the HN relationship. The fitting results of the HN equation showed that the asymmetry of the dielectric loss peak β increases with decreasing concentration levels of La x. For a severe dielectric relaxation (for example, the significant decrease of the k-value with increasing frequency for the $La_{0.09}Zr_{0.91}O_{2-\delta}$ thin film), the width change of the loss peak α played an important role during data fitting. For the $La_{0.09}Zr_{0.91}O_{2-\delta}$ thin film, it was found that the combined CS+KWW relaxation process ($\beta_k = 0.53$ and n = 0.91) can be substituted by the HN function where distribution parameters α and β were 0.6535 and 0.3458, respectively because both the HN function and the combined CS+KWW relationship had two distribution parameters.

It was found that dielectric relaxation is related to the size of the crystal grains formed during annealing and that doping affects the size of the crystal grains formed.

ACKNOWLEDGEMENT

This research was funded in part from the National Natural and Science Foundation of China under the grant No. 60976075 and from the Suzhou Science and Technology Bureau of China under the grant SYG201007.

REFERENCES

1.   Juan, P.C.; Liu, C.H.; Lin, C.L.; Ju, S.C.; Chen, M.G.; Chang, I.Y.K.; Lu, J.H. Electrical characterization and dielectric property of MIS capacitors using a high-k $CeZrO_4$ ternary oxide as the gate dielectric. Jpn. J. Appl. Phys. 2009, 48, 1–5.

2.   Dong, G.F.; Qiu, Y. Pentacene thin-film transistors with Ta_2O_5 as the gate dielectric. J. Korean Phys. Soc. 2009, 54, 493–497.

3.   Zhu, X.H.; Zhu, J.M.; Li, A.D.; Liu, Z.G.; Ming, N.B. Challenges in atomic-scale characterization of high-k dielectrics and metal gate electrodes for advanced CMOS gate stacks. J. Mater. Sci. Technol. 2009, 25, 289–313.

4.   International Technology Roadmap for Semiconductors Home page. Available online: http://public.itrs.net/ (accessed on 17 May 2012).

5.   Ricco, B.; Olivo, P.; Nguyen, T.N.; Kuan, T.S.; Ferriani, G. Oxide-thickness determination in thin-insulator MOS structures. IEEE Trans. Electron Devices 1988, 35, 432–438.

6.   Terman, L.M. An investigation of surface states at a silicon/silicon oxide interface employing metal-oxide-silicon diodes. Solid State Electron. 1962, 5, 285–299.

7.   Lee, S.W. A capacitance-based method for experimental determination of metallurgical channel length of submicron LDD MOSFET's. IEEE Trans. Electron Devices 1964, 41, 403–412.

8.   Huang, C.L.; Faricelli, J.V.; Arora, N.D. A new technique for measuring MOSFET inversion layer mobility. IEEE Trans. Electron Devices 1993, 40, 1134–1139.

9.   Sze, S.M. Physics of Semiconductor Devices; Wiley: New York, NY, USA, 1981.

10. Yang, K.J.; Chenming, H. MOS capacitance measurements for high-leakage thin dielectrics. IEEE Trans. Electron Devices 1999, 46, 1500–1501.

11. Choi, C.H.; Wu, Y.; Goo, J.S.; Yu, Z.; Dutton, R.W. Capacitance reconstruction from measured C-V in high leakage nitride/oxide MOS. IEEE Trans. Electron Devices 2000, 47, 1843–1850.

12. Nicollian, E.H.; Brews, J.R. MOS (metal oxide semiconductor) physics and technology. In MOS (Metal Oxide Semiconductor) Physics and Technology; Wiley: NewYork, NY, USA, 2003.

13. Kwa, K.S.K.; Chattopadhyay, S.; Jankovic, N.D.; Olsen, S.H.; Driscoll, L.S.; O'Neill, A.G. A model for capacitance reconstruction from measured lossy MOS capacitance-voltage characteristics. Semicond. Sci. Technol. 2003, 18, 82–87.

14. Hirose, M.; Hiroshima, M.; Yasaka, T.; Miyazaki, S. Characterization of silicon surface microroughness and tunneling transport through ultrathin gate oxide. J. Vac. Sci. Technol. A 1994, 12, 1864–1868.

15. Schuegraf, K.F.; King, C.C.; Hu, C. Impact of polysilicon depeletion in thin oxide MOS technology. In International Symposium on VLSI Technology, Systems and Applications-VLSI-TSA, Taipei, Taiwan, 12–14 May 2012; VLSI Technology: Seattle, WA, USA; pp. 86–90.

16. Lee, S.W.; Liang, C.; Pan, C.S.; Lin, W.; Mark, J.B. A study on the physical mechanism in the recovery of gate capacitance to C_{ox} in implant polysilicon MOS structure. IEEE Electron Device Lett. 1992, 13, 2–4.

17. Spinelli, A.S.; Pacelli, A.; Lacaita, A.L. An improved formula for the determination of the polysilicon doping. IEEE Electron Device Lett. 2001, 22, 281–283.

18. Hauser, J.R.; Ahmed, K. Characterization of ultra-thin oxides using electrical C-V and I-V measurements. In AIP Conference Proceedings, Gaithersburg, MD, USA, November 1998; AIP: College Park, MD, USA; pp. 235–239.

19. Pregaldiny, F.; Lallement, C.; Mathiot, D. Accounting for quantum mechanical effects from accumulation to inversion, in a fully analytical surface potential-based MOSFET model. Solid State Electron. 2004, 48, 781–787.

20. Sune, J.; Olivo, P.; Ricco, B. Quantum-mechanical modeling of accumulation layers in MOS structure. IEEE Trans. Electron Devices 1992, 39, 1732–1739.

21. Pregaldiny, F.; Lallement, C.; van Langevelde, R.; Mathiot, D. An advanced explicit surface potential model physically accounting for the quantization effects in deep-submicron. Solid State Electron. 2004, 48, 427–435.

22. Satter, M.M.; Haque, A. Modeling effects of interface trap states on the gate C-V characteristics of MOS devices on alternative high-mobility. Solid State Electron. 2010, 54, 621–627.

23. jiong, L.Z.; Ma, T.P. A new method to extract EOT of ultrathin gate dielectric with high leakage current. IEEE Electron Device Lett. 2004, 25, 655–657.

24. Bierwagen, O.; Geelhaar, L.; Gay, X.; Piešinš, M.; Riechert, H.; Jobst, B.; Rucki, A. Leakage currents at crystallites in $ZrAl_xO_y$ thin films measured by conductive atomic-force microscopy. Appl. Phys. Lett. 2007, 90, 232901:1–232901:3.

25. Böscke, T.S.; Govindarajan, S.; Kirsch, P.D.; Hung, P.Y.; Krug, C.; Lee, B.H. Stabilization of higher-tetragonal HfO_2 by SiO_2 admixture enabling thermally stable metal-insulator-metal capacitors. Appl. Phys. Lett. 2007, 91, 072902:1–072902:3.

26. Zhao, C.Z.; Taylor, S.; Werner, M.; Chalker, P.R.; Murray, R.T.; Gaskell, J.M.; Jones, A.C. Dielectric relaxation of lanthanum doped zirconium oxide. J. Appl. Phys. 2009, 105, 044102:1–044102:8.

27. Gaskell, J.M.; Jones, A.C.; Aspinall, H.C.; Przybylak, S.; Chalker, P.R.; Black, K.; Davies, H.O.; Taechakumput, P.; Taylor, S.; Critchlow, G.W. Liquid injection ALD and MOCVD of lanthanum aluminate using a bimetallic alkoxide precursor. J. Mater. Chem. 2006, 26, 3854–3860.

28. Niwa, M.; Kouzaki, T.; Okada, K.; Udagawa, M.; Sinclair, R. Atomic-order planarization of ultrathin SiO_2/Si (001) interfaces. Jpn. J. Appl. Phys. 1994, 33, 388–394.

29. Shiau, J.J.; Fahrenbruch, A.L.; Bube, R.H. Influence of ac contact impedance on high-frequency, low-temperature, or fast-transient junction measurements in semiconductors. J. Appl. Phys. 1987, 61, 1556–1561.

30. Venkatesan, V.; Das, K.; von-Windheim, J.A.; Geis, M.W. Effect of back contact impedance on frequency dependence of capacitance-voltage measurements on metal/diamond diodes. Appl. Phys. Lett. 1993, 63, 1065–1067.

31. Lo, S.H.; Buchanan, D.A.; Taur, Y.; Wang, W. Quantum-mechanical modeling of electron tunneling current from the inversion layer of ultra-thin-oxide nMOSFET's. IEEE Electron Device Lett. 1997, 18, 209–211.

32. Taechakumput, P.; Zhao, C.Z.; Taylor, S.; Werner, M.; Pham, N.; Chalker, P.R.; Murray, R.T.; Gaskell, J.M.; Aspinall, H.C.; Jones, A.C. Origin of frequency of dispersion in high-k dielectrics. In Proceedings of 7th International Semiconductor Technology Conference ISTC2008,

Shanghai, China, 15–17 March 2008; Electrochemical Society-Asia: Hong Kong, China; pp. 20–26.

33. Zhao, X.; Vanderbilt, D. First-principles study of structural vibrational and lattice dielectric properties of hafnium oxide. Phys. Rev. B 2002, 65, 1–4.

34. Wu, W.H.; Tsui, B.Y.; Huang, Y.P.; Hsieh, F.C.; Chen, M.C.; Hou, Y.T.; Jin, Y.; Tao, H.J.; Chen, S.C.; Liang, M.S. Two-frequency C-V correction using five-element circuit model for high-k gate dielectric and ultrathin oxide. IEEE Electron Device Lett. 2006, 27, 399–401.

35. Lerner, E.J. The end of the road for Moore's law. IBM J. Res. Develop. 1999, 6, 6–11.

36. Ahmed, K.; Ibok, E.; Yeap, G.C.F.; Qi, X.; Ogle, B.; Wortman, J.J.; Hauser, J.R. Impact of tunnel currents and channel resistance on the characterization of channel inversion layer charge and polysilicon-gate depletion of sub-20-A gate oxide MOSFET's. IEEE Trans. Electron Devices 1999, 46, 1650–1655.

37. Choi, C.H.; Jung-Suk, G.; Tae-Young, O.; Yu, Z.P.; Dutton, R.W.; Bayoumi, A.; Min, C.; Voorde, P.V.; Vook, D.; Diaz, C.H. MOS C-V characterization of ultrathin gate oxide thickness (1.3–1.8 nm). IEEE Electron Device Lett. 1999, 20, 292–294.

38. Zhang, J.L.; Yuan, J.S.; Ma, Y.; Oates, A.S. Modeling of direct tunneling and surface roughness on C-V charateristics of ultra-thin gate MOS capacitors. Solid State Electron. 2001, 45, 373–377.

39. Arora, N.D.; Rios, R.; Huang, C.L. Modeling the polysilicon depletion effect and its impact on submicrometer CMOS circuit performance. IEEE Trans. Electron Devices 1995, 42, 935–943.

40. Rios, R.; Arora, N.D.; Huang, C.L. An analytic polysilicon depletion effect model for MOSFET's. IEEE Electron Device Lett. 1994, 15, 129–131.

41. Gupta, A.; Fang, P.; Song, M.; Lin, M.R.; Wollesen, D.; Chen, K.; Hu, C. Accurate determination of ultrathin gate oxide thickness and effective polysilicon doping of CMOS devices. IEEE Electron Device Lett. 1997, 18, 580–582.

42. Takeuchi, K.; Kasai, N.; Terada, K. A new effective channel length determination method for LDD MOSFETs. In IEEE 1991 International Conference on Microelectronic Test Structure, Kyoto, Japan, 18–20 March 1991; IEEE: New York, NY, USA, 1991; pp. 215–220.

43. Cumberbatch, E.; Abebe, H.; Morris, H.; Tyree, V. Aanalytical surface potential model with polysilicon gate depletion effect for NMOS. In

Proceedings of 2005 Nanotechnology conference, Anaheim, CA, USA, 8–12 May 2005; CRC Press: Boca Raton, FL, USA; pp. 57–60.

44. Abebe, H.; Cumberbatch, E.; Morris, H.; Tyree, V. Numerical and analytical results for the polysilicon gate depletion effect on MOS gate capacitance. In Proceedings University/Government/Industry Microelectronics Symposium, San Jose, CA, USA, 25–28 June 2006; IEEE: New York, NY, USA; pp. 113–117.

45. Lin, W.W. A simple method for extracting average doping concentration in the polysilicon and silicon surface layer near the oxide in polysilicon-gate MOS structures. IEEE Electron Device Lett. 1994, 15, 51–53.

46. Ohkura, Y. Quantum effects in Sin-MOS inversion layer at high substrate concentration. Solid State Electron. 1990, 33, 1581–1585.

47. Yeo, Y.C.; Ranade, P.; King, T.J.; Hu, C. Effects of high-k gate dielectric materials on metal and silicon gate work function. IEEE Electron Device Lett. 2002, 23, 342–344.

48. Sue, J.; Olivo, P.; Riccb, B. Quantum-mechanical modeling of accumulation layers in MOS structure. IEEE Electron Device Lett. 1992, 39, 1732–1739.

49. Moglestue, C. Self-consistent calculation of electron and hole inversion charges at silicon-silicon dioxide interfaces. J. Appl. Phys. 1986, 59(5), 3175–3183.

50. Richter, C.A.; Hefner, A.R.; Vogel, E.M. A comparison of quantum-mechanical capacitance-voltage simulators. IEEE Electron Device Lett. 2001, 22, 35–37.

51. Debye, P. Polar Molecules; Chemical Catalogue Company: New York, NY, USA, 1929.

52. Cole, K.S.; Cole, R.H. Dispersion and absorption in dielectrics. J. Chem. Phys. 1941, 9, 341–351.

53. Davidson, D.W.; Cole, R.H. Dielectric relaxation in glycerol, propylene glycol and n-propanol. J. Chem. Phys. 1951, 19, 1484–1490.

54. Raju, G.G. Dielectrics in Electric Fields; CRC Press: Boca Raton, FL, USA, 2003.

55. Davidson, D.W.; Cole, R.H. Dielectric relaxation in glycerine. J. Chem. Phys. 1950, 18.

56. Bello, A.; Laredo, E. Distribution of relaxation times from dielectric spectroscopy using Monte Carlo simulated annealing: Application to α-PVDF. Phys. Rev. B 1999, 60, 12764–12774.

57. Kliem, H.; Arlt, G. A relation between dielectric distribution functions and structural properties of amorphous matter.Annu. Rep. Conf. Electr. Insul. Dielectr. Phenom. 1987, 56, 325.

58. Cabeza, M.; Keddam, M.; Novoa, X.R.; Sanchez, I.; Takenouti, H. Impedance spectroscopy to characterize the pore structure during the hardening process of Portland cement paste. Electrochim. Acta 2006, 51, 1831–1841.

59. Havriliak, S.; Negami, S. A complex plane analysis of α-dispersions in some polymer systems. J. Polym. Sci. Pt. C 1966,14, 99–117.

60. Havriliak, S.; Negami, S. A complex plane representation of dielectric mechanical relaxation processes in some polymers. Polymer 1967, 8, 161–210.

61. Kalgaonkar, R.A.; Nandi, S.; Tambe, S.S.; Jog, J.P. Analysis of viscoelastic behavior and dynamic mechanical relaxation of copolyester based layered silicate nanocomposites using Havriliak-Negami model. J. Polym. Sci. B Polym. Phys. 2004, 42, 2657–2666.

62. Hartmann, B.; Lee, G.F.; Lee, J.D. Loss factor height and width limits for polymer relaxations. J. Acoust. Soc. Am. 1994,95(1), 226–233.

63. Jonscher, A.K. Dielectric Relaxation in Solids; Chelsea Dielectric Press: London, UK, 1983.

64. Williams, G.; Watts, D.C. Non-symmetrical dielectric relaxation behaviour arising from a simple empirical decay function. Trans. Faraday Soc. 1969, 66, 80–85.

65. Bokov, A.A.; Ye, Z.G. Double freezing of dielectric response in relaxor $Pb(Mg_{1/3}Nb_{2/3})O_3$ crystals. Phys. Rev. B 2006,74.

66. Ngai, K.L.; Plazek, D.J. A quantitative explanation of the difference in the temperature dependences of the viscoelastic softening and terminal dispersions of linear amorphous polymers. J. Polym. Sci. Polym. Phys. 1986, 24, 619–632.

67. Rendell, R.W.; Ngai, K.L.; Rajagopal, A.K. Volume recovery near the glass transition temperature in poly(vinyl acetate): predictions of a coupling model. Macromolecules 1987, 20, 1070–1083.

68. Ngai, K.L.; Fytas, G. Interpretation of differences in temperature and pressure dependences of density and concentration fluctuations in amorphous poly(phenylmethyl siloxane). J. Polym. Sci. Polym. Phys. 1986, 24, 16833–1694.

69. Yoshihara, M.; Work, R.N. Dielectric relaxation in undiluted poly(4-chlorostyrene). II. Characteristics of the high-frequency tail. J. Chem. Phys. 1981, 74, 5872–5876.

70. Shioya, Y.; Mashimo, S. Comparison between interpretations of dielectric behavior of poly (vinyl acetate) by the coupling model and the Havriliak-Negami equation. J. Chem. Phys. 1987, 87, 3173–3177.

71. Boese, D.; Kremer, F.; Fetters, F.J. Molecular dynamics in bulk cis-polyisoprene as studied by dielectric spectroscopy. Macromolecules 1990, 23, 829–835.

72. Boese, D.; Momper, B.; Meier, G.; Kremer, F.; Hagenah, J.U.; Fischer, E.W. Molecular dynamics in poly(methylphenylsiloxane) as studied by dielectric relaxation spectroscopy and quasielastic light scattering. Macromolecules 1989, 22, 4416–4421.

73. Alvarez, F.; Alegria, A. Relationship between the time-domain Kohlrausch-Williams-Watts and frequency-domain Havriliak-Negami relaxation functions. Phys. Rev. B 1991, 44, 7306–7312.

74. Bokov, A.A.; Kumar, M.M.; Xu, Z.; Ye, Z.G. Non-arrhenius stretched exponential dielectric relaxation in antiferromagnetic $TiBO_3$ single crystals. Phys. Rev. B 2001, 64.

75. Curie, J. Recherches sur le pouvoir inducteur specifique et sur la conductibilite des corps cristallises. Ann. Chim. Phys.1889, 18, 385–434.

76. Von Schweidler, E. Studien uber die anomalien im verhalten der dielektrika. Ann. Phys. 1907, 24, 711–770.

77. Lee, B.; Moon, T.; Kim, T.; Choi, D. Dielectric relaxation of atomic-layer-deposited HfO_2 thin films from 1 kHz to 5 GHz. Appl. Phys. Lett. 2005, 87, 012901:1–012901:3.

78. Costa, L.C.; Henry, F. Dielectric universal law of lead silicate glasses doped with neodymium oxide. J. Non-Cryst. Solids 2007, 353, 4380–4383.

79. Hoerman, B.H.; Ford, G.M.; Wessels, B.W. Dielectric properties of epitaxial $BaTiO_3$ thin films. Appl. Phys. Lett. 1998,73(16), 2248–2250.

80. Horikawa, T.; Makita, T.; Kuriowa, T.; Mikami, N. Dielectric Relaxation of (Ba, Sr)TiO_3 thin films. Jpn. J. Appl. Phys.1995, 34.

81. Werner, M.; Zhao, C.Z.; Taylor, S.; Chalker, P.R.; Potter, R.J.; Gaskell, J. Permittivity enhancement and dielectric relaxation of doped hafnium and zirconium oxide. In IEEE 16th International Symposium on the Physical and Failure Analysis of Integrated Circuits, Suzhou, China, 6–10 July 2009; IEEE: New York, NY, USA; pp. 625–627.

82. Fischer, D.; Kersch, A. The effect of dopants on the dielectric constant of HfO_2 and ZrO_2 from first principles. Appl. Phys. Lett. 2008, 92, 012908:1–012908:3.

83. Dervos, C.T.; Thirios, E.; Novacovich, J.; Vassiliou, P.; Skafidas, P. Permittivity properties of thermally treated TiO_2.Mater. Lett. 2004, 58, 1502–1507.

84. Choosuwan, H.; Guo, R.; Bhalla, A.S.; Balachandran, U. Low-temperature dielectric behavior of Nb_2O_5-SiO_2 solid solutions. J. Appl. Phys. 2003, 93, 2876–2879.

85. Yu, H.; Liu, H.; Hao, H.; Guo, L.; Jin, C.; Yu, Z.; Cao, M. Grain size dependence of relaxor in $CaCu_3Ti_4O_{12}$ ceramics.Appl. Phys. Lett. 2007, 91, 222911:1–222911:3.

86. Sivakumar, N.; Narayanasamy, A.; Chinnasamy, C.N.; Jeyadevan, B. Influence of thermal annealing on the dielectric properties and electrical relaxation behavior in nanostructured $CoFe_2O_4$. J. Phys. Condens. Matter 2007, 19.

87. Zhao, C.Z.; Werner, M.; Taylor, S.; Chalker, P.R.; Jones, A.C.; Zhao, C. Dielectric relaxation of La-doped Zirconia caused by annealing ambient. Nanoscale Res. Lett. 2011, 6, 1–6.

Chapter 8

APPLICATION OF HFCT AND UHF SENSORS IN ON-LINE PARTIAL DISCHARGE MEASUREMENTS FOR INSULATION DIAGNOSIS OF HIGH VOLTAGE EQUIPMENT

Fernando Álvarez [1], Fernando Garnacho [2], Javier Ortego [3] and Miguel Ángel Sánchez-Urán [1]

[1]Department of Electrical Engineering, Polytechnic University of Madrid, Ronda de Valencia 3, Madrid 28012, Spain

[2]LCOE-FFII, Eric Kandel 1, Getafe 28906, Spain

[3]DIAEL, Peñuelas 38, Madrid 28005, Spain

ABSTRACT

Partial discharge (PD) measurements provide valuable information for assessing the condition of high voltage (HV) insulation systems, contributing to their quality assurance. Different PD measuring techniques have been developed in the last years specially designed to perform on-line measurements. Non-conventional PD methods operating in high frequency bands are usually used when this type of tests are carried out. In PD measurements the signal acquisition, the subsequent signal processing and the capability to obtain an accurate diagnosis are conditioned by the selection of a suitable detection technique and by the implementation of effective signal processing tools. This paper proposes an optimized electromagnetic detection method based on the combined use of wideband PD sensors for measurements performed in the HF and UHF frequency ranges, together with the implementation of powerful processing tools. The effectiveness of the measuring techniques proposed is demonstrated through an example, where several PD sources are measured simultaneously in a HV installation consisting of a cable system connected by a plug-in terminal to a gas insulated substation (GIS) compartment.

INTRODUCTION

The analysis of the types of defects and the degradation modes in different insulation materials of HV electrical systems has shown that the presence of PD is a very common characteristic in all of them [1,2,3]. Once PD activity is detected the identification and location of the associated type of defect is very important to evaluate whether the discharges are harmful or not.

On-line PD measurements have become a common practice for assessing the insulation condition of installed HV equipment. This type of testing is carried out during normal operation of the electrical system. For utilities, the most interesting advantage of on-line PD measurements is that once the sensors are installed in the power grid, the electrical service is not interrupted for the measurements. Another advantage of on-line tests is that PD activity can be acquired under different load conditions in temporal or permanent monitoring applications, which results very useful for the identification of certain types of defects and allows the analysis of the defects' evolution over time.

Among the different PD measuring techniques available (electrical, acoustic, optical and analysis of chemical byproducts) [4,5,6,7,8], the electrical method is the most widely used because of its effectiveness. Despite the usefulness of this technique, the presence of electric noise disturbances when on-line PD tests are performed represents a drawback due to the loss of sensitivity, especially when PD pulses with low energy are present in the HV installation. Another drawback of on-line PD measurements appears when more than one source of pulse-shaped signals is present in the electric HV system under testing. In these cases, an adequate selection of the non-conventional measuring technique with the most suitable sensors and the implementation of effective signal processing tools are essential to achieve a correct evaluation of the insulation condition. Different approaches have been developed and applied to PD measurement and processing techniques in order to achieve accurate diagnoses. Most of them are focused on wideband measurements with different types of sensors [5,9,10], noise rejection techniques [11], pulse classification processes [12,13,14] and defect location and identification [15,16]. In the study presented herein a complete measuring and processing method providing a comprehensive solution for the most common problems arising in on-line measurements is described.

The advantages of the complementary measurements applying non-conventional methods measuring with high frequency current transformers (HFCT) and UHF sensors are analyzed. These measurements performed with a developed measuring device integrated with powerful processing tools enable to perform better diagnosis and thus to improve the electrical system reliability.

In Section 2 the technical details and practical implementations of HFCT and UHF sensors for their application in complementary PD measurements are analysed and the sensors used in this study are described. In this section a UHF-HF converter designed to measure pulses captured in the UHF range using PD instruments for the acquisition of signals in the HF range is also described. In the following section the measuring instrument developed is described, together with three signal processing tools specially designed for on-line PD tests. Finally, experimental PD measurements applying HF and UHF sensors are performed according with a proposed measuring procedure and the results obtained are analyzed in detail.

PD SENSORS AND FREQUENCY CONVERTER FOR PD MEASUREMENTS IN UHF

In the detection of PD pulses by electromagnetic methods, two techniques are mainly distinguished, those that apply the conventional method based in the standard IEC 60270, in which PD pulses are measured in a frequency range below 1 MHz and those that implement non-conventional methods based in the use of sensors measuring in the HF (3–30 MHz), VHF (30–300 MHz) and UHF (300 MHz–3 GHz) frequency ranges [4,17,18]. A new technical specification (IEC 62478) which also covers non-conventional electrical methods in addition to the acoustical technique is currently under consideration. It is expected that this technical specification will be published in 2015.

The conventional method is used as reference for the quality assurance of the elements of HV electrical systems. This method, while suitable for laboratory measurements, is not appropriate for on-line measurements, since the noise conditions in most cases are very high for the measuring frequencies specified in the standard (≤ 1 MHz). To achieve an appropriate sensitivity in on-line measurements and to analyze in detail the captured signals, non-conventional methods with larger frequency ranges and bandwidths than those used according to the IEC 60270 standard are applied.

Moreover, to perform the measurements applying the conventional method, a coupling capacitor in series with a measuring impedance are used. In non-conventional methods the PD signals are captured with the sensors coupled directly in the HV equipment and the external coupling capacitor is not required, which suppose another advantage for on-line tests.

During the last three decades, the use of non-conventional methods implemented with HF and UHF sensors for on-line PD measurements has been generalized in order to detect incipient insulation defects. Theoretical and practical aspects to be known about these sensors are considered in this section.

Furthermore, different measuring devices have been designed to perform non-conventional PD tests. A measuring instrument has been developed by the authors together with a UHF-HF converter to measure the signals captured with both UHF and HF sensors. The frequency converter connected to the output of UHF sensors allows processing the signals acquired with less costly acquisition cards used to measure in the HF range. The UHF-HF converter developed and the advantages of its implementation are described in the last point of this section.

HFCT Sensors

HFCT sensors are widely used for PD detection and their application for the location and identification of PD sources is very effective. A HFCT sensor also called radio frequency current transducer (RFCT) consists of an induction coil with a ferromagnetic core suitable for the measurement of transient signals as PD or pulse-shaped noise interferences. In general, when on-line PD measurements are performed on HV installations, HFCT sensors are clamped in the grounding conductors of the earthing network. For this application, the sensor can be modeled as a system in which the input is the current of the PD pulses flowing through it and the output is the induced voltage that is measured over the input impedance of the measuring instrument (usually 50 Ω), see Figure 1.

The transfer function of these magnetic sensors $V = f(B)$ can be expressed by the Faraday's law of induction.

$$e = -n \cdot \frac{d\Phi}{dt} = -n \cdot A \cdot \frac{dB}{dt} = -\mu_0 \cdot n \cdot A \cdot \frac{dH}{dt} \tag{1}$$

where Φ is the magnetic flux passing through the coil of the secondary side which is formed by a number of turns n and presents an area A. In the case of a coil with a ferromagnetic core, Equation (1) can be written as Equation (2):

$$e = -\mu_0 \cdot \mu_r \cdot n \cdot A \cdot \frac{dH}{dt} \tag{2}$$

The induced voltage in the secondary is proportional to the rate of change of current in the primary, being the mutual inductance between the earth conductor and the secondary M, the proportional constant.

$$e = M \cdot \frac{di}{dt}$$

(3)

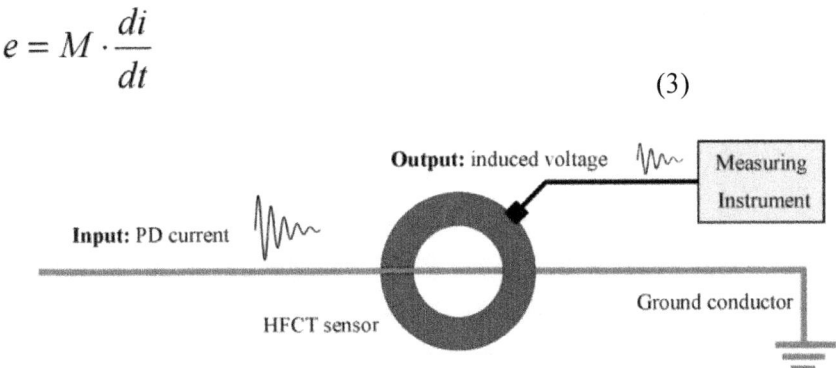

Figure 1: HFCT sensor placed in a ground conductor for PD measurement.

Air-coil sensors present linearity for large frequency bandwidths; however the sensitivity of the sensor improves significantly when soft magnetic materials are used for the transformer core. The ferromagnetic core introduces to the transfer function of the sensor some nonlinear factors which depend mainly on the frequency but also on the temperature and flux density. For this reason this type of sensors should be designed according with a specified frequency characteristic [19].

In the origin of the discharge the current pulses generated have rise times of a few units of nanosecond or even below 1 nanosecond, presenting a frequency spectrum with significant components up to hundreds of MHz or even units of GHz [18,20,21,22]. The high frequency components of the PD signals are significantly filtered when the pulse propagates to the earth conductors where the HFCT sensors are placed. For the detection and location of PD events in electrical grids only the frequency components below 10 MHz are of interest [23].

Figure 2b shows the frequency spectrum of a PD pulse measured in different positions along a power cable system using an on-line monitoring system. The pulse was generated in an insulation defect of a cable joint in the position CB1 and was captured by the HFCT sensors located in the positions CB1 (0 m), CB2 (675 m), CB3 (1732) and GIS-1(1060), see Figure 2a. The bandwidth of the sensors used is 100 kHz–25 MHz and the sampling rate of the acquisition board is 100 MS/s. It was observed an attenuation for the upper frequencies with the increasing distance, due to the behavior of the propagation path as a low-pass filter for PD pulses. Furthermore, the frequency content of the recorded pulses was always below 20 MHz. Considering this analysis and in order to comply with appropriate requirements of sensitivity adopting a good economical solution, the use of HFCT sensors with a bandwidth from hundreds of kHz to 20 MHz is recommended for their practical implementation in PD measurements.

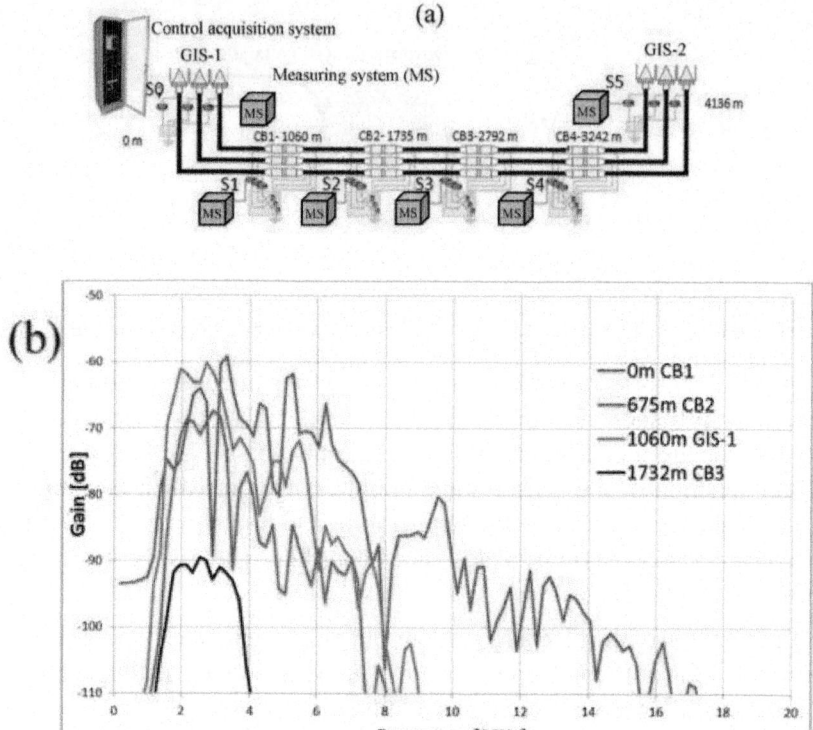

Figure 2: (a) On-line PD measurement with a monitoring application in a 45 kV XLPE power cable system; (b) Frequency spectrum of a PD pulse measured in different positions.

PD measurements in the HF range using HFCT sensors provide considerable advantages:

- The sensitivity is not so pulse shape dependent as in conventional PD measuring instruments.

- The signal to noise ratio (SNR) can be improved, analyzing the data in certain frequency bands.

- High sensitivity is obtained when the sensors are located closed to the PD source and also when they are far from it. In a power cable system, when a PD pulse propagates through the cable shield, although the high frequency content of the signal is filtered, the pulse can be measured at distances exceeding one kilometer holding their spectral content up to units of megahertz [23].

- If two or more HFCT sensors are placed in a HV installation, the measurement of the PD pulses with a common time reference allows the determination of the location of defects by the time-of-flight analysis.

- PD pulses waveform can be recorded for post processing purposes. The recorded signals can be classified by the characterization of the pulse shape with the aim of discriminate different PD or noise sources. A proper classification of the recorded pulses and a subsequent analysis of the associated phase resolved PD (PRPD) patterns, improves the sensibility in the detection of defects and facilitates more accurate diagnoses.

- For the frequency range specified, ferrite cores are commonly available and a high quality HFCT sensor manufacture results easy and inexpensive.

In this study, a clamp-type HFCT sensor has been designed for the measurement of PD in the ground connections of HV equipment in electrical power installations, see Figure 3a. Figure 3b shows the frequency response of the sensor, which presents a bandwidth from 100 kHz to 20 MHz.

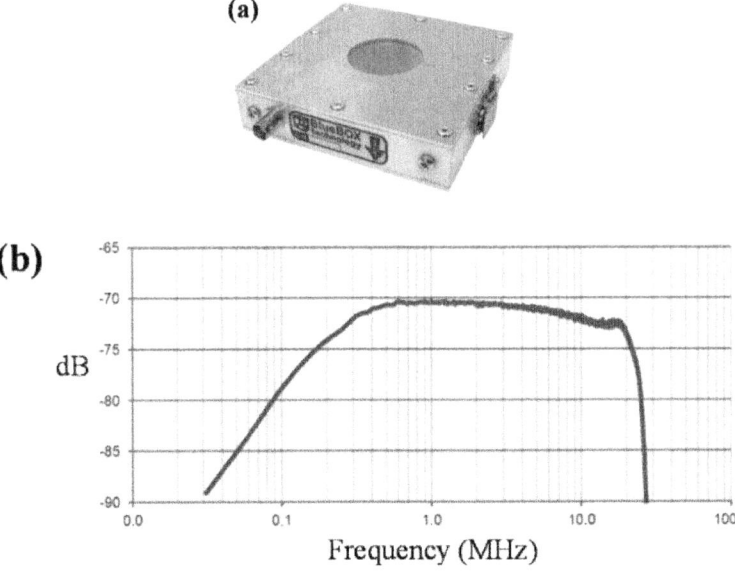

Figure 3. (a) Clamp type HFCT sensor; (b) Frequency response.

UHF Sensors

Although UHF sensors for PD measurements are considered antennas, they do not measure in the far-field but rather in the near-field, where the power of the electromagnetic waves of PD signals is sufficiently significant. UHF sensors can be classified as internal or external couplers, according to whether they are mounted inside or outside the HV equipment. Internal UHF sensors are normally assembled during construction in the enclosure of GIS compartments, occasionally in power transformers and they can also be embedded in cable accessories. External UHF sensors consist of portable couplers that are fitted in inspection windows or in exposed barrier edges in GIS, transformers and other HV elements such as metal enclosed switchgear or rotating machines. External sensors can also be coupled on the sheath of power cables and on their accessories.

Several studies have been conducted to analyze the benefits of the UHF technique with invasive and external sensors, especially in the resonant metallic chambers of GIS and in power transformers [4,24].

Although wire antennas (dipoles and monopoles), aperture antennas (horn type) and array antennas (log-periodic) are used for PD measurements [25,26,27], the most common UHF sensors for this application are microstrip antennas based on patch type couplers; these consist mainly on a disc or a printed planar resonant structure [28,29,30,31]. This study is focused in the implementation of patch antennas for PD measurements.

A patch coupler presents a high immunity to the noise environment because it has a high directivity and measures only EM-wave propagating in one direction. This is an advantage when PD measurements are performed with non-invasive sensors. Patch sensors for measurements in the UHF range consists of a conducting plate above a ground plane, an example of this configuration is shown in Figure 4. For the disc resonant structure presented, Stratton's treatment for the TM wave is applied to the co-ordinate system indicated [32]. Provided that the disc is close to the ground plane in comparison with its radius, two dimensional resonances are considered and the angular resonant frequencies ω_{nm} are given by:

$$\omega_{nm} = \frac{\chi_{nm}}{a \cdot \sqrt{\mu \cdot \varepsilon}}$$

(4)

where a is the metal plate radius, μ the permeability of the dielectric, ε the permittivity and χ_{nm} the mth nonzero root of $J_n'(x) = 0$, being $J_n'(x)$ the derivative with respect to x of the Bessel function $J_n(x)$ of the first kind of order n [29,32,33].

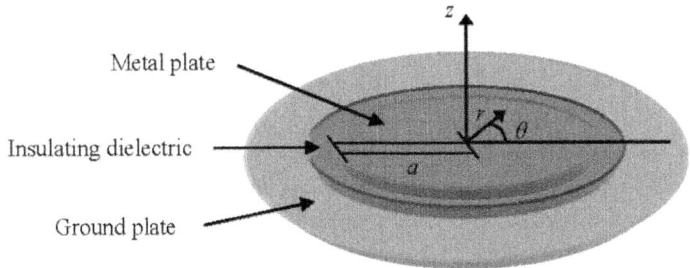

Figure 4: Resonant disc structure for UHF patch sensors.

The electric field component E_z under the disc at resonance can be expressed with the polar coordinates r and θ:

$$E_z = k \cdot \sin(\omega_{nm} \cdot t) \cdot J_n \cdot \left(\frac{r}{a} \cdot \chi_{nm} \right) \cdot \cos(n\theta)$$

(5)

where k is a constant.

For design considerations the resonant frequencies are calculated using Equation (4) and the electric field amplitude E_z under the disc for the resonant modes is determined by Equation (5).

Although patch type antennas are resonant in nature and are mainly used in telecommunication applications matched to a particular resonant frequency with a high gain, they can also be designed to match a larger frequency range by increasing the bandwidth of resonance frequencies to be used for PD measurements in a UHF range. Moreover, even though the sensor is not well matched for bandwidths out of the resonance frequencies, PD signals can be also measured with a sufficient gain in these frequency bands. A detailed explanation for the construction and characterization aspects of resonance disc UHF antennas is presented in [29,34]. Furthermore, improvements in the design of this type of UHF sensors to reduce the effect of the coupler on the detected PD spectrum have been presented in the studies developed by [29,30]. Other designs of patch type sensors for PD measurements have been implemented with a different geometrical shape of the patch conductor [21], or with a different configuration of the resonant structure [35].

As was mentioned before PD pulses are very fast, with rise times even below 1 nanosecond, so at their origin they radiate electromagnetic waves with spectral content that can extends to frequencies of 1 GHz or higher. When the discharges are generated inside a GIS compartment or are confined in the enclosure of a power transformer, an electrical resonance is produced that causes the pulses to have a duration of up to 1 μs. Due to the characteristics of the insulating medium that these elements present, the inherent losses of the pulses are very low, for example, in the case of a GIS in the presence of barriers and discontinuities the attenuation is 2 dB/m. However, when the insulation medium is solid, as in power cables, this behaves as a low pass filter and the UHF components of the pulses are significantly attenuated with the distance from the focus [36]. This characteristic is useful to be selective in the location of PD sources when UHF measurements are performed in cables and accessories; if PD pulses are detected in the UHF range the sensor is close to the origin of the defect.

An experimental measurement was performed in order to characterize the attenuation of the UHF frequency components of PD pulses propagating in power cables. The experimental setup consisted of a 12/20 kV XLPE cable of 500 m length, see Figure 5. A UHF calibrator (ref. LDIC LDC-5/UHF) was used to inject pulses in the power cable through one of its terminals. The pulses were measured every 0.5 m by means of a UHF non-invasive sensor with a bandwidth from 20 to 800 MHz and a digital oscilloscope of 1 GHz bandwidth and 5 GS/s of sample rate. The measuring frequency range was delimited from 300 MHz to 800 MHz using a band pass filter at the output of the sensor.

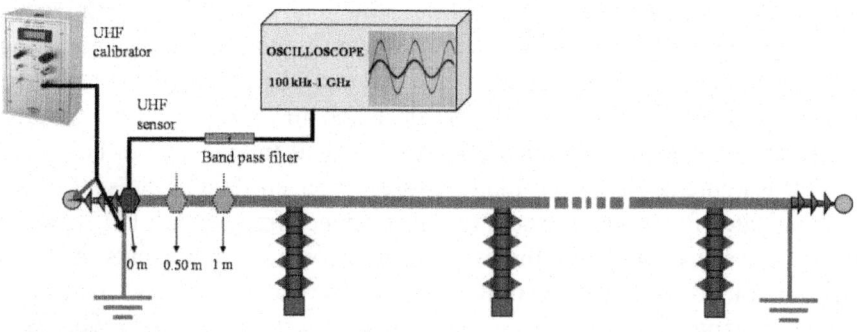

Figure 5: Experimental setup for the evaluation of the UHF frequency components of a PD pulse in a power cable.

The spatial selectivity for PD location is characterized by a decreasing exponential curve as shown in Figure 6a. In the first 2 m the pulse magnitude is attenuated by 67%. At a distance of 5 m no pulse can be detected with the acquisition system. Figure 6b shows the superposition of a pulse measured at 0 m and at 3.5 m (yellow and red signals respectively). It can be observed that the amplitude decreases and that the shape of the signal changes with the distance covered. Comparing the frequency spectrum of both signals, the amplitude values obtained for the frequency range from 600 to 800 MHz when the sensor is positioned at the beginning of the cable are negligible when the sensor is at 3.5 m, see Figure 6c,d. This characterization confirms that the location of defects in power cables can be performed in a very selective way.

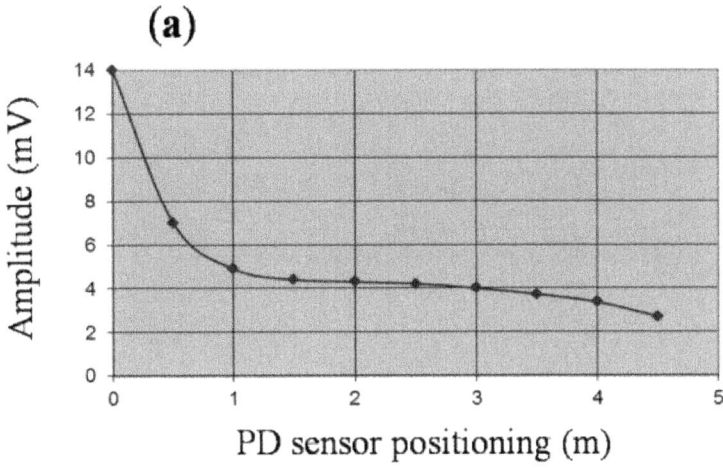

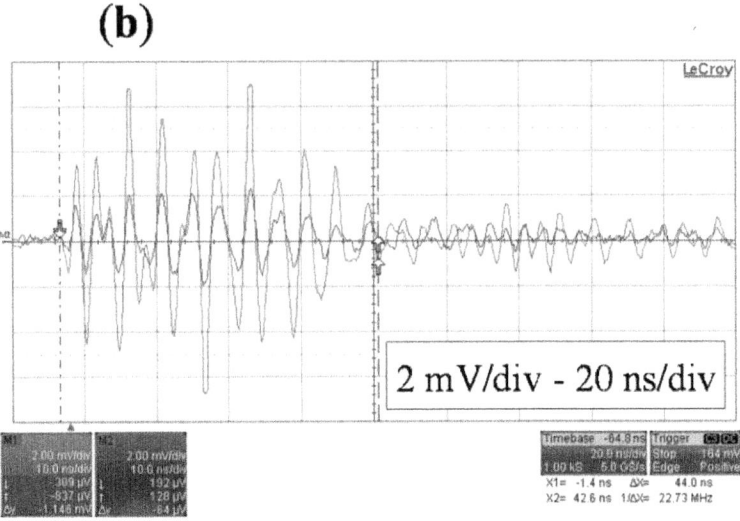

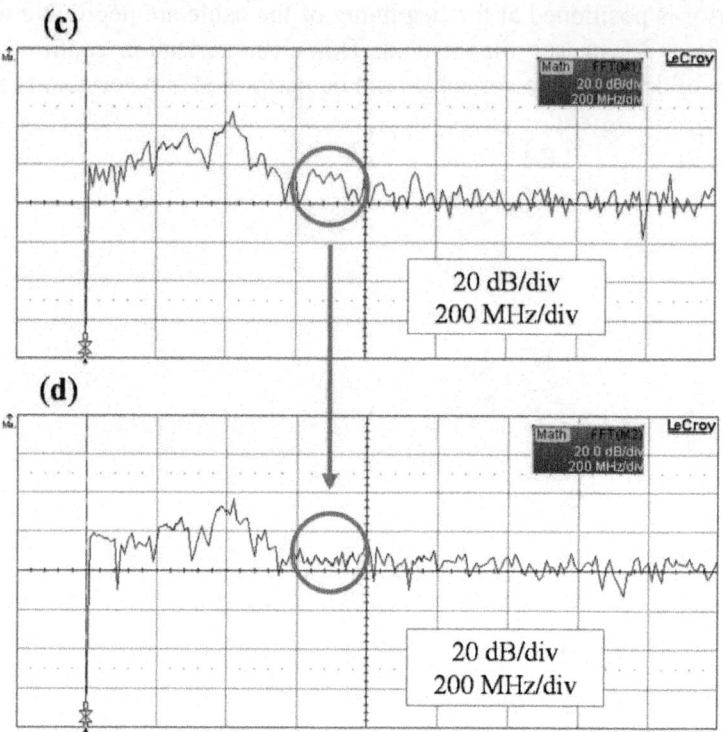

Figure 6: (a) PD pulse attenuation with the distance; (b) Superposition of a pulse measured at the beginning of the cable and at 3.5 m; (c) Frequency spectrum of the pulse measured at 0 m; (d) Frequency spectrum of the pulse measured at 3.5 m.

The following advantages are obtained when PD measurements are performed with UHF sensors.

- High immunity to electric noise, interferences and corona discharges in air, since the frequency spectrum of these signals in the UHF range is very low and in some emplacements negligible.
- High sensitivity in PD detection inside shielded GIS compartments, metal enclosed switchgears and transformer tanks, due to the inner electrical resonance, very low inherent losses and low noise levels.
- Possibility of PD source location. In the case of GIS and transformers the location is performed using several sensors and analyzing the arrival times of the pulses (time-of-flight measurements).
- Accurate defect locations are achieved in cables and accessories due to the selectivity in the distance that this technique presents.

- Possibility to discriminate between internal and external defects to the HV equipment under consideration.

In the study presented two types of commercial UHF patch sensors were used to perform the measurements together with the UHF-HF converter developed. The first one is an invasive disc coupler specially design for GIS compartments and metal enclosed switchgears, with a bandwidth form 0.1 to 1.5 GHz and a load impedance recommended of 50 Ω (ref. Siemens 926 98850 174 B). Figure 7 shows the metal plate of the sensor and a view of its installation in a GIS compartment.

Figure 7: (a) Metal plate of the invasive UHF sensor; (b) External view of the sensor installed in a GIS compartment.

The other type of patch sensor is a non-invasive antenna (ref. IPEC OSM CCT2) designed to be coupled in the exposed barrier edges of the metal cladding of GIS, switchgear cabinets or power transformers and on the surface of power cables and their accessories, see Figure 8. It has a bandwidth of 20–800 MHz and its recommended load impedance is 50 Ω.

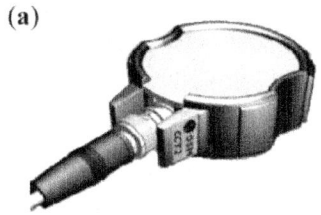

(b)

Figure 8: (a) UHF non-invasive patch coupler; (b) Sensor installed on the surface of a power cable.

For PD measurements in UHF, where the noise level is very low, the use of an amplifier connected to the output of the sensors is recommended, since when low amplitude pulses are captured, this element helps to achieve more sensitivity when they are recorded by the measuring equipment. Furthermore, when the pulses present an amplitude slightly above the noise level, the use of an amplifier helps to differentiate in more detail a possible PRPD pattern. In this study, the PD measurements with UHF sensors have been performed with an amplifier of 20 dB. This amplifier has a bandwidth from 100 kHz to 1 GHz and an input impedance of 50 Ω.

The measurements with the UHF non-invasive sensor have been performed delimiting its bandwidth in a defined frequency range starting from 300 MHz. By doing this, a greater immunity to noise and interference signals of low and high frequency is obtained, being the signal to noise ratio (SNR) higher. Furthermore, as was explained before, PD measurements in cable accessories from 300 MHz allow more selectivity in the location of defects. Considering the importance of these two advantages, a high pass filter of 300 MHz was connected at the output of the non-invasive UHF sensor. Besides, in order to avoid radiofrequency disturbances mainly generated by telecommunication applications it is recommendable to implement also a low pass filter of 800 MHz.

Figure 9a shows the configuration of the UHF invasive sensor with the amplifier and the UHF-HF converter connected to its output. The final configuration of the non-invasive sensor coupled on a cable with the band-pass filter (300–800 MHz), the amplifier and the converter is shown in Figure 9b.

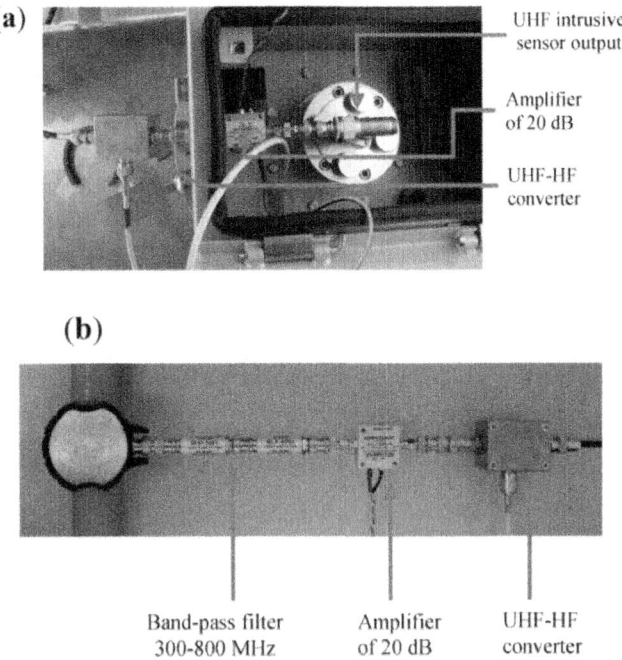

(a)
UHF intrusive
sensor output

Amplifier
of 20 dB

UHF-HF
converter

(b)

Band-pass filter   Amplifier   UHF-HF
300-800 MHz   of 20 dB   converter

Figure 9: (a) Outside view of the UHF invasive sensor with amplifier and UHF-HF converter; (b) Non-invasive sensor with band-pass filter, amplifier and converter.

UHF-HF Converter

Processing PD signals measured in the UHF frequency range requires oscilloscopes with a bandwidth of at least 1 GHz and a 5 GS/s sampling rate. These oscilloscopes are very expensive and are not provided with a specific software designed to record and analyze the raw data acquired with UHF sensors.

Commercial PD instruments designed to measure in the HF range with a sampling rate around 100 MS/s, can be used to record and process the measured UHF signals after they are transformed into HF signals by means of a frequency converter; these instruments are less expensive than the oscilloscope specified to measure in UHF. The UHF-HF converter developed in this study enables to take advantages of capturing the pulses in UHF and in turn record and analyze the signals with a measuring instrument designed to perform PD measurements with HF sensors.

The design of the converter is based on an integrator model that presents a delay in the output signal of less than 40 ns. The converter detects the peak of the UHF input signals and applies a level shift to convert the signals to continuous levels. Furthermore, it is integrated with an internal amplifier which gain can be adjusted. Its bandwidth is from 0.1 to 3 GHz.Figure 10 shows a pulse measured in UHF and the same signal converted in a HF pulse. It can be observed that the original signal has significant values of amplitude in the frequency spectrum up to 500 MHz, while the output signal of the converter presents significant values up to 5 MHz.

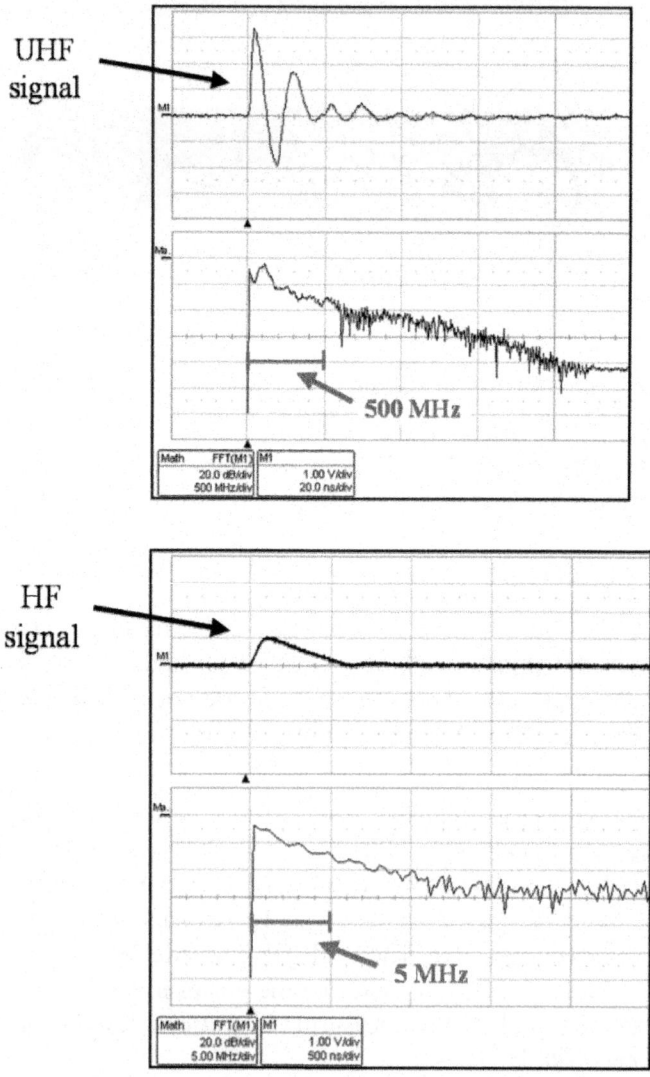

Figure 10: Frequency analysis of an original UHF pulse and the transformed signal.

MEASURING INSTRUMENT AND SIGNAL PROCESSING TOOLS

Measuring Instrument

A measuring device was designed to perform PD measurements in on-line conditions. It is equipped with an acquisition board of 14 bits of vertical resolution, a bandwidth of 50 MHz and a sampling frequency of 100 MS/s. The signal processing tools described below were developed and integrated in the processing and analysis stages of the PD instrument in order to improve the ability to perform accurate diagnosis. The equipment is controlled by a computer connected via fiber optic cable or through a wireless connection. The front view of a measuring unit and the display screen for PD acquisition and post processing is shown in Figure 11. To be measured by this measuring instrument, pulses captured with UHF sensors must be converted to HF pulses with frequency content at most 50 MHz. This conversion is performed with the UHF-HF converter described in previous section.

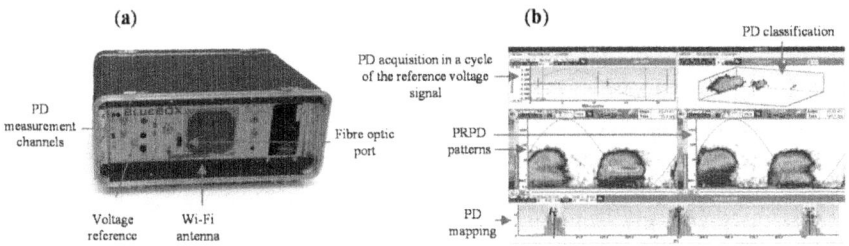

Figure 11: (a) Measuring instrument; (b) Display of an acquisition in a PD test.

Signal Processing Tools

The effectiveness of the new developments and applications of DP processing tools has increased the use of PD measurements for maintenance purposes. By implementing certain processing tools, an accurate assessment of the HV elements insulation condition can be achieved. Three processing tools have been designed and implemented in the measuring instrument developed: noise filtering, automatic PD location and PD sources classification [37].

Noise Filtering Tool

A filtering tool based on the wavelet transform (WT) makes it possible to discriminate pulse signals from continuous background noise. Much effort has been focused on de-noising and detecting transient signals implementing wavelets algorithms [38,39]. The filtering tool developed is based in the use of the WT, together with a statistical analysis that evaluates the standard deviation of the background noise level to discriminate the existence of PD activity. Applying the implemented wavelet filter, pulse-shaped signals with amplitudes even below the background noise level can be detected, see Figure 12.

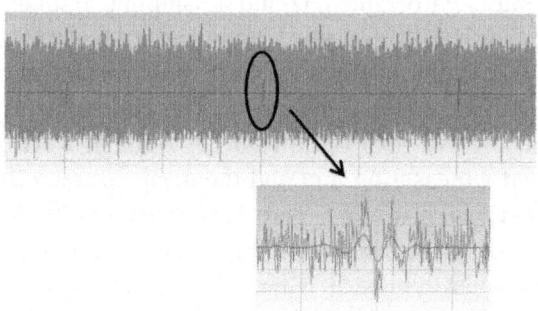

Figure 12: Result of the filtering tool. Discrimination of pulse-shaped signals overlaid with background noise.

The signals presenting a transient behavior (PD and pulse-type interferences) are filtered and selected for further processing. The most significant difference with traditional PD instruments is that for the pulse selection is not necessary to set any noise threshold level; all those pulses filtered by signal processing are recorded.

After the filtering process not all the selected signals correspond to PD pulses as it is not possible to discriminate pulse-type disturbances presenting a transient behavior similar to PD. Nevertheless, applying the classification by location tool and the sources classification tool enables distinguishing clusters associated to PRPD patterns corresponding with this type of noise sources.

Classification by Location Tool

When a pulse source is present in a HV equipment the signals generated propagate through the grid in different directions according to the impedance characteristics of the medium. As an example, in the case of a cable system two pulses travel towards opposite directions along the cable length from the source site. Each pulse arrives in a certain instant t_i to a PD sensor connected to a measuring unit, see Figure 13.

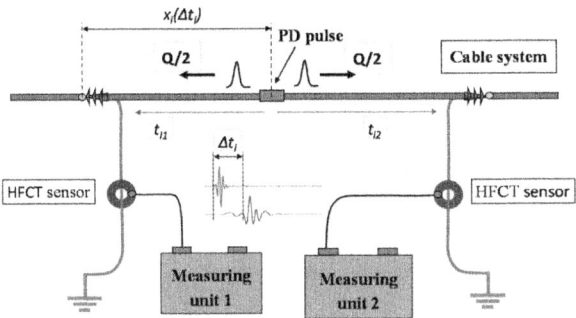

Figure 13: Time delay between the arrival times to the measuring units.

The synchronization of different measuring units for PD location can be accomplished in different ways: by sending a synchronizing signal through a fiber optic cable, by the time reference of the coordinated universal time (CUT) from the pulse per second signal (PPS) of a global positioning system receiver (GPS), or by a reference signal sent through the installation under test from a measuring unit to the others. In this study the synchronization by GPS has been used. The PPS is generated by a GPS module that achieves an accuracy within 15 ns (1σ).

A pulse recorded in a measuring unit is correlated with another pulse recorded by other unit only if the time delay between the arrival times of both pulses Δt_i is less than the propagation time between sensors t_w defined in Equation (6):

$$t_w = L / V \tag{6}$$

being L the distance between sensors and V the propagation speed of the signals.

By determining the time delay Δt_i for the correlated pulses captured by each sensor, the measuring system is able to establish automatically the location of the different pulse sources. On the basis of the knowledge of the propagation time t_w and the cable length L, the location of a pulse source $x_i(\Delta t_i)$ for the correlated pulses is established by the following expression:

$$x_i\left(\Delta t_i\right) = \frac{L}{2} \cdot \left[1 - \left(\frac{\Delta t_i}{t_w} \right) \right] \tag{7}$$

where:

$$\Delta t_i = t_{i2} - t_{i1} \tag{8}$$

All correlated pulses are positioned in a PD mapping diagram and the different locations of pulse sources are identified (see Figure 14).

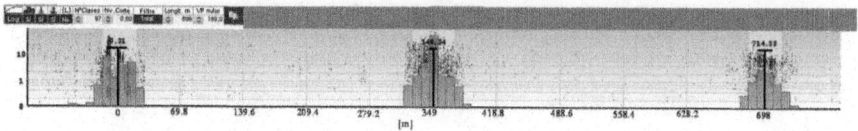

Figure 14: Example of pulse classification by location. PD sources located in different sites of a cable system.

PD clustering Tool

When more than one PD source is present in a HV installation and especially if their emplacement is in the same position it is necessary to differentiate and to identify them. In this processing tool, a damped oscillating wave $f_i(t)$ is associated to each filtered pulse, that is modeled by means of a sine function $g_i(t)$ and modulated by an enveloping function $h_i(t)$ that fits the local maxima of the absolute values of the signal as shown in Figure 15a:

$$f_i(t) = g_i(t) \cdot h_i(t)$$

(9)

$$g_i(t) = sin(w_i \cdot t - \varphi_i)$$

(10)

$$h_i(t) = \frac{A_i}{e^{\alpha_i(t)} + e^{-\beta_i(t)}}$$

(11)

The parameter $w_i = 2\pi f_i$ for $g_i(t)$ is estimated using the Fourier transform of the filtered signal and corresponds with the maximum amplitude value of the spectrum. The shape of the enveloping function $h_i(t)$ is mainly described by the damping coefficients of the exponential terms α and β, which are related with the steepness in the increasing and decreasing intervals, respectively; the parameter A_i characterizes the amplitude of the envelope. By adjusting the values of α, β and A_i it is possible to obtain a suitable enveloping function to modulate the sinusoidal waveform. The criterion to select the best fitting for the function $h_i(t)$ is the least squares. The parameters of the modeled pulses f_p, α_i and β_i characterize their waveform providing useful information for the classification by clusters of the pulses originated in different sources of the HV installation. Implementing the three dimensional diagram shown in Figure 15b using these waveform parameters, different pulse sources can be distinguished by selecting the formed clusters.

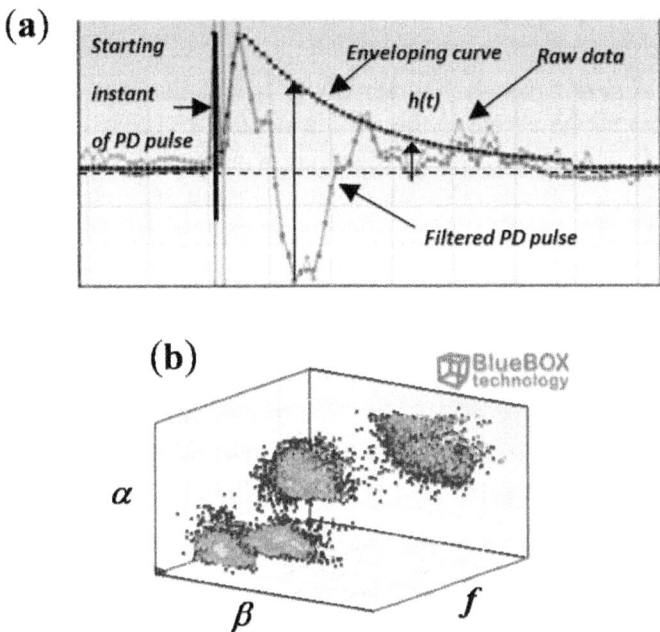

Figure 15: (a) Enveloping function for a filtered PD pulse; (b) 3D diagram for the classification of pulse sources by clusters.

HF AND UHF PD MEASUREMENT

In order to analyze the advantages of the complementary measurement with HF and UHF sensors applying suitable processing and analysis tools, four types of insulation defects were generated and measured simultaneously in a laboratory experimental setup. The PD sources implemented were: two internal defects in XLPE, corona in air and corona in SF_6 inside the metal cladding of a GIS compartment.

Generation of PD Sources

Internal discharges (voids inside solid dielectrics) are considered the most harmful for dielectric elements and lead to insulation failures. Corona discharges pose no risk if they are generated in air, however, if they are produced close to silicone insulators or in SF_6, the consequences also lead to insulation faults.

Internal defect type 1. A cavity type defect was caused in a cable termination at the end of the semiconducting external layer. The cavity was done by a transversal cut of 1.5 mm depth in the main insulation (XLPE), as shown in Figure 16a. This type of defect can be caused due to an incorrect handling of the splicing tools at the edges of semiconducting shield cut-backs, in accessories assembling processes [20].

Internal defect type 2. Another cavity type defect was caused in a cable termination making a void in the main insulation (XLPE). A hole of 1.5 mm depth was drilled using a 1 mm drill bit removing previously a part of the cable semiconducting external layer, as shown in Figure 16b. The semiconducting layer was fixed again to the main insulator and the cuts were reconstructed using a semiconducting varnish (ref. Raychem EPPA 220). This type of defect occurs due to failures in the cable extrusion process that causes a void in the main insulation, or when a void remains inside the insulation elements after the assembly process of the cable accessories.

Corona in Air (Tip as HV Electrode): Corona discharges were caused by means of a point-plane configuration assembled in a holding device that allows the adjustment of the gap between the point and the plane electrodes (see Figure 16c).

Corona in a GIS Compartment (Tip as Ground Electrode): To simulate a corona defect in SF_6, a 1 mm radius tip was fixed in the enclosure of a GIS compartment that was manufactured with a stainless steel metal cladding of 3 m length. This chamber has a thickness of 3 mm and an external diameter of 500 mm. A conductor of 54 mm diameter was assembled inside the compartment, see Figure 17b. The internal conductor was supported by two epoxy resin cone spacers. The metal cladding is provided with two inspection windows available for the assembling of invasive PD sensors as shown in Figure 17a.

The mechanical and electrical connection of the tip with the metal cladding was performed using neodymium magnets, which provides a reliable joining and prevents the machining of the structure, see Figure 16d. The gap between the tip and the internal conductor can be adjusted with a screw. For a gas pressure inside the compartment of 4 bars and applying a test voltage of 20 kV, the distance between the tip and the internal conductor was adjusted in 3 mm. Figure 16d shows the tip assembled inside the GIS chamber. In practice, corona discharges in SF_6 can be caused by protrusions due to errors in the installation process, to the friction between moving parts, to burrs or to material degradations caused by switching operations during the service life of the GIS [40].

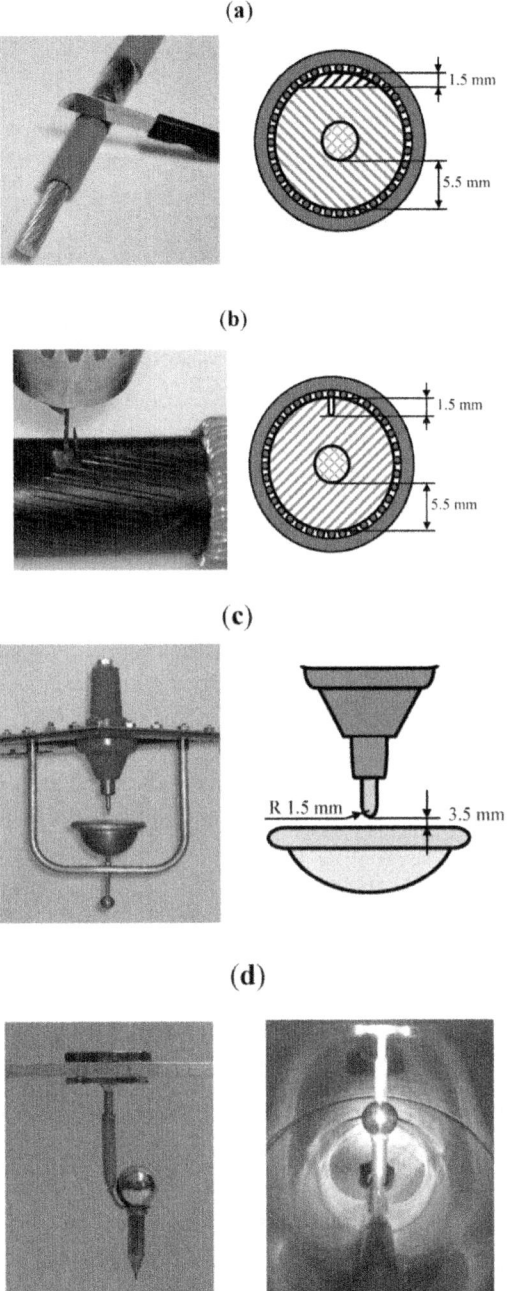

Figure 16: PD sources. (a) Transversal cut in the main insulation of a cable termination. (b) Internal void inside the main insulation of a cable; (c) Point-plane corona in air; (d) Corona in a GIS chamber.

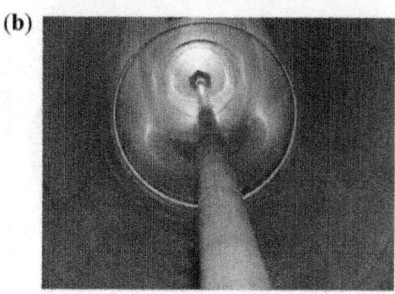

Figure 17: (a) Outside view of the GIS compartment; (b) Interior view.

Experimental Setup

An experimental arrangement implemented in a HV laboratory was configured in order to simulate a representative HV installation and to perform significant DP measurements with the proposed sensors. This installation was composed by the following subsystems, see Figure 18.

(a) **Substation:** Formed by the GIS compartment described above and a 12/20 kV XLPE cable of 15 m length and a 240 mm^2 aluminum conductor. This cable interconnects by a plug-in terminal (position B) and a cable junction (position C) the GIS compartment with the cable system.

(b) **Cable system:** A 12/20 kV XLPE cable system with a 240 mm^2 aluminum conductor was configured with a total length of 867 m. The line was composed by two joined cable coils with lengths of 350 and 517 m. The connection between them (position D) was made with a splice.

(c) **Distribution grid:** To simulate the continuity of the distribution grid a 12/20 kV XLPE cable system with a total length of 585 m and a 240 mm^2 aluminum conductor was connected to the cable system. The connection (position E) was made simulating a junction of the switchgear cabinets of a distribution transformer substation.

A resonant generator of 36 kV with a 22.9 H reactor was used to apply a test voltage of 51 Hz and 20 kV. This voltage level was above the inception voltage of each defect and provided stable measurements. The HV generator was connected between the cable system and the connecting cable with the GIS chamber (position C). In this point the synchronization reference signal for the PD measurements was acquired by a capacitive divider.

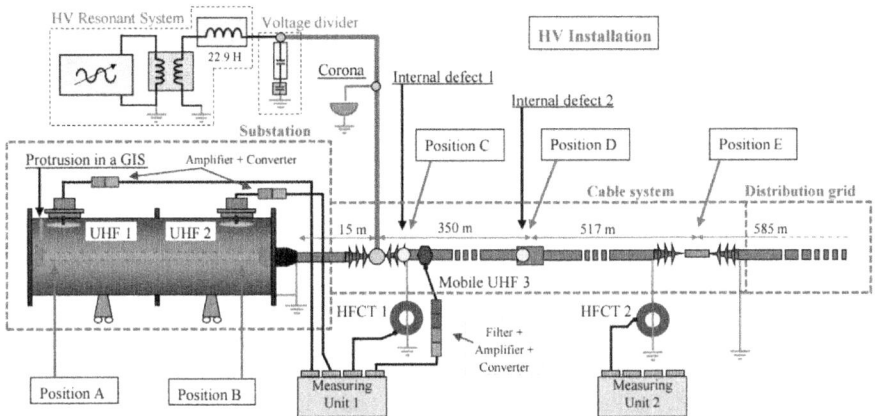

Figure 18: Experimental setup for the measurement of insulation defects with HF and UHF sensors.

The four insulation defects described above were measured simultaneously in the HV installation arranged simulating an on-line test. These defects and their position in the test object are specified in Table 1.

Table 1: Type of defect and location

Insulation Defect	Location
1. Corona in a GIS compartment	A (inside the GIS chamber)
2. Internal defect type 1 in a cable termination	C (in the power cable)
3. Corona in air	C (in the power cable)
4. Internal defect type 2 in one of the cable terminations of a splice	D (in the power cable)

Measuring Procedure

In order to be effective enough in the detection and identification of defects, in a first step the measurements were performed with two HFCT sensors for the cable system and with two UHF invasive sensors for the GIS compartment. The measurements with these sensors placed in strategic positions, performed with the measuring instrument described inSection 3.1, present a good technical and economic solution for the complete evaluation of the HV installation.

Due to the attenuation and dispersion effects of PD signals in power cables until they are captured by the HFCT sensors, the maximum distance between sensors recommended is around 1 km. This distance can be increased to 2 km at the expense of a loss of sensitivity that is compensated by the noise filtering technique applied. In order to gain a general overview of the entire setup proposed, the HFCT sensors were installed in the ground connections of the cable system in positions C and E (see Figure 18), being the distance between them 867 m.

By coupling invasive UHF sensors in the GIS compartment, defects in its interior can be detected and located with high sensitivity. To provide a proper coverage of the substation in terms of sensitivity when PD measurements are performed in GIS, it is recommended to couple a UHF sensor approximately every 15–20 m. Due to the dimensions of the GIS chamber implemented in the experimental setup, the two invasive UHF sensors were placed inside the compartment with a distance between them of only 2 m.

In a second step, if any insulation defect is detected with the HFCT sensors, a mobile non-invasive UHF sensor can be coupled in different sites of the HV installation to determine the exact position of the PD source. A summary of the sensors and their position in the test object is presented in Table 2.

Table 2: Type of sensors and location in the test setup

Invasive UHF Sensors	Position
UHF 1	A
UHF 2	B
Non-invasive HFCT Sensors	**Position**
HFCT 1	C
HFCT 2	E
Non-invasive UHF Sensor	**Position**
Mobile UHF 3	In different emplacements along the HV installation to determine the exact PD source location

The PD measurements were performed with two measuring instruments of four acquisition channels. Table 3 indicates the connection of the sensors to the measuring units.

Table 3: Measuring instrument and channel assigned to each sensor

Measuring Instrument	Channel	Sensor
Measuring Unit 1	Channel 1	Invasive UHF 1 (position A)
	Channel 2	Invasive UHF 2 (position B)
	Channel 3	HFCT 1 (position C)
	Channel 4	Mobile non-invasive UHF 3
Measuring Unit 2	Channel 1	HFCT 2 (position E)
	Rest of channels	Free for more sensors if required

Experimental Results and Analysis

The performance of the described sensors and processing tools is validated by the measurement of the PD sources in the presence of noise interferences. A significant amount of pulses were processed in order to obtain representative patterns for the individual sources. In practice, a thorough PD analysis is required, since up to four defects are present in the HV installation, with the added difficulty of the emplacement of two of them in the same site (position C).

Firstly the measurements performed with the HFCT sensors are analyzed. The PRPD patterns for the complete acquisition obtained with both sensors after the application of the wavelet noise filter are shown in Figure 19. As the individual patterns generated in each source are mixed, the insulation condition of the HV system can not be determined in a first assessment.

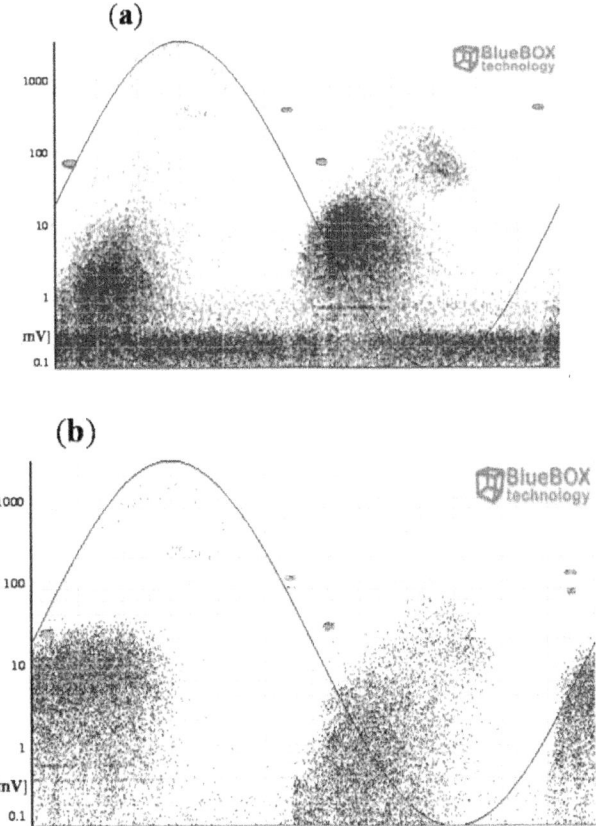

Figure 19: PRPD patterns obtained with the HFCT sensors. (a) HFCT 1 in position C; (b) HFCT 2 in position E.

A further analysis is required for an accurate identification of the defects. In a first step the pulse-type noise sources and interferences measured in the HV installation are removed. Pulse-shaped electric noise can be differentiated from PD pulses with the PD classification tool by pulse waveform. The results of this processing tool are shown in Figure 20. For the measurement performed with the sensors HFCT 1 and HFCT 2, 6 and 5 clusters have been differentiated respectively. Analyzing the PRPD patterns for each group of pulses it has been possible to correlate all the clusters obtained with each sensor except the cluster 6 corresponding to the sensor HFCT 1.

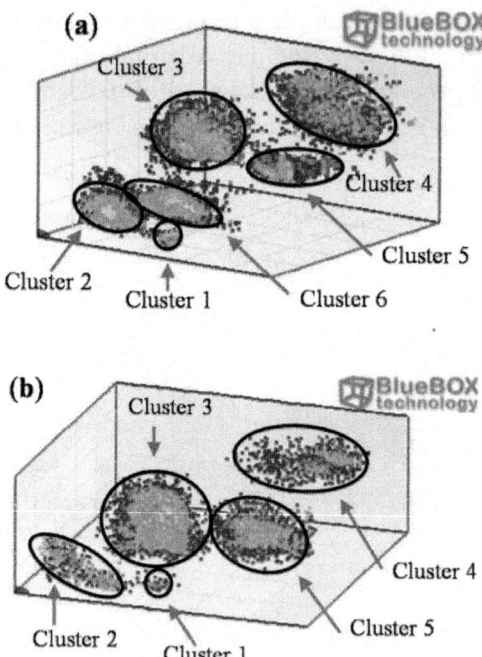

Figure 20: Result of the PD classification tool for the acquisitions with HFCT sensors 1 (a) and 2 (b).

By selecting the pulses corresponding to cluster 1, the electronic noise generated by the power electronic (IGBTs) of the resonant system used to apply high voltage is identified. These interferences are synchronized with the test voltage reference signal. Furthermore, a significant amount of random pulse-shaped signals conducted by the earth network of the installation were also identified by selecting the pulses of cluster 2. The PRPD patterns corresponding to clusters 1 and 2 are shown in Figure 21a,b. Once these noise signals are removed, the remaining pulses considered PD, are shown in Figure 21c,d.

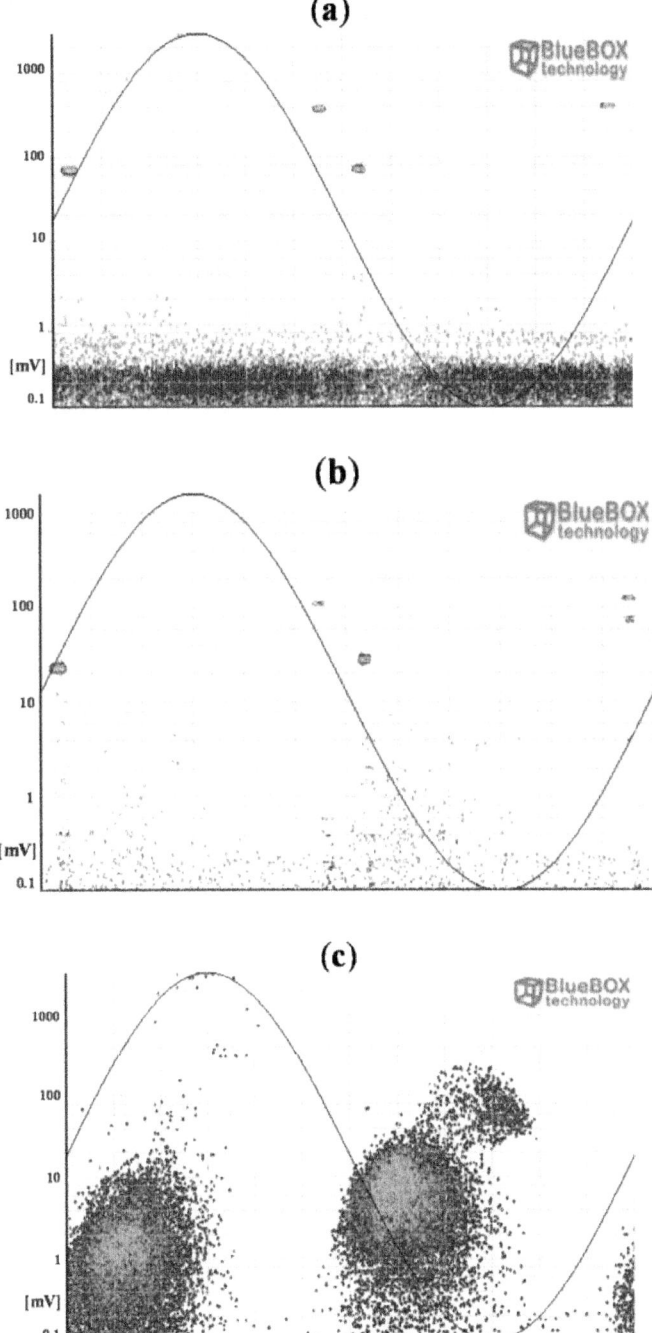

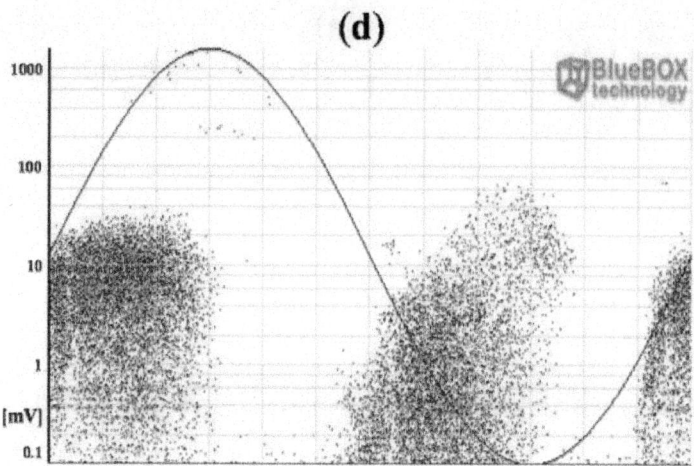

Figure 21: PRPD patterns of the noise pulses measured with the HFCT sensor 1 (a) and HFCT sensor 2 (b) corresponding to clusters 1 and 2. PRPD patterns of the PD pulses measured with the HFCT sensor 1 (c) and 2 (d).

In a second step, the classification by location tool is applied to the PD pulses and the PD mapping shown in Figure 22 is obtained. Analyzing the time delay Δt_i for correlated pulses, two different emplacements of PD sources were detected (positions C and D in Figure 18 and Figure 22). These sites correspond with the emplacement of the sensor HFCT 1 and with the joint in the cable system respectively.

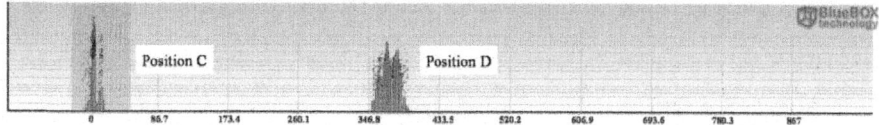

Figure 22: PD classification by location. PD sources located in positions C and D of the cable system.

In a third step, the classification tool by pulse waveform is applied to those pulses corresponding with a specific location. In position C two clusters have been identified for the measurements performed with sensor 1, see Figure 23. These two groups of pulses correspond with clusters 3 and 4 in Figure 20a. The individual PRPD patterns are obtained by selecting the formed clusters, see Figure 23b,c.

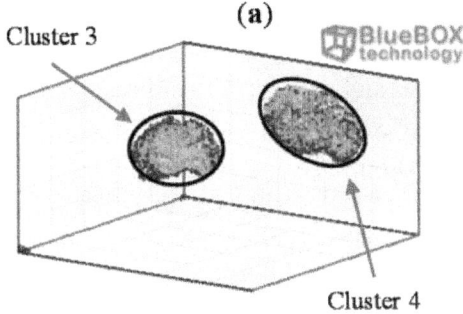

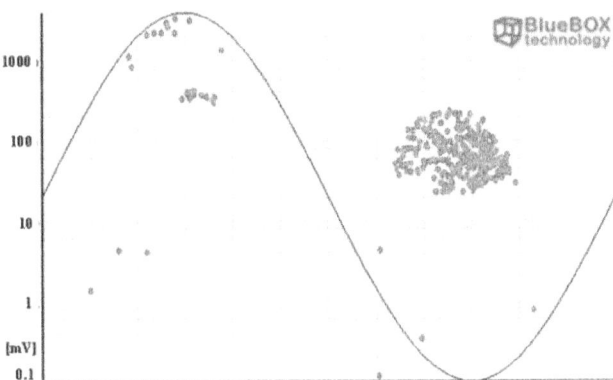

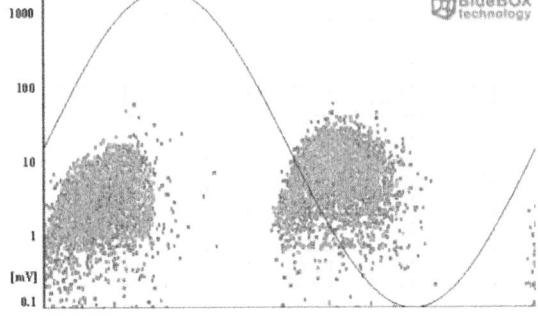

Figure 23: (a) Classification of PD pulses positioned in C, (b) and (c) PRPD patterns for the sources positioned in C.

The pattern displayed for cluster 3 is characteristic of a corona defect (tip as HV electrode). PD pulses appear on the crests of the reference voltage signal. There are stable amplitude values within a certain range in both half-periods and there are less pulses and with higher amplitude in the positive half-cycle.

The pattern corresponding to cluster 4 is characteristic of an internal defect in a solid dielectric. PD pulses occur slightly before the zero crossings and in the increasing intervals of the voltage signal. There is certain symmetry and a similar repetition rate when comparing the patterns of both half-periods.

In position D only the group of pulses corresponding to cluster 5 was identified (see Figure 20a and Figure 24a). The pattern obtained is also characteristic of an internal defect, see Figure 24b. The patterns shown for positions C and D were obtained for the signals acquired by the sensor HFCT 1.

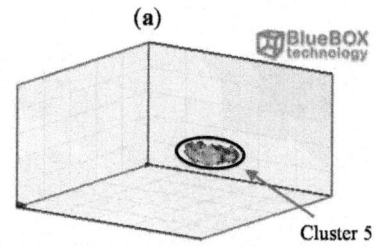

(a)

Cluster 5

(b) PD pattern for Cluster

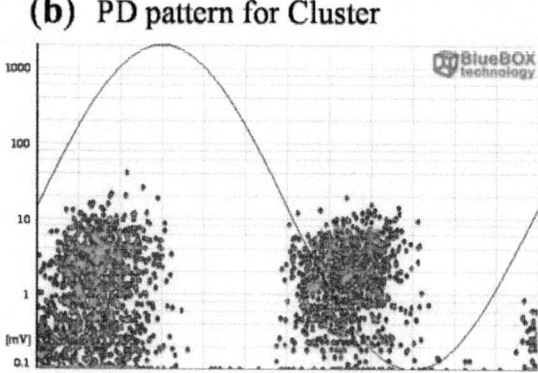

Figure 24: (a) Classification of PD pulses positioned in D; (b) PRPD pattern for the source positioned in D.

In order to corroborate the location of the defects in the HV installation, it must be considered that:

- When the pulses are positioned by the location tool in the same site where a HFCT sensor is coupled the location of the defect can not be totally assured, *i.e.*, the source can be in that position or in a previous one. This is because in both cases the delay in the arrival time of the pulses to the sensors is the same, see Figure 25a,b, so the pulses are always positioned in the emplacement where the HFCT sensor is placed.

- Only when the defect is in an intermediate point between the sensors it can be assured that the positioning of the focus is correct; a certain time delay corresponds only to one emplacement of the PD source, see Figure 25c.

According with this and considering the internal defects indicated in the PD mapping it can be assured only the correct emplacement for the defect of position D (positioned between the sensors).

The emplacement of the corona defect in position C can also be verified, as it is only in this point where the high voltage conductor of the cable system is exposed to air.

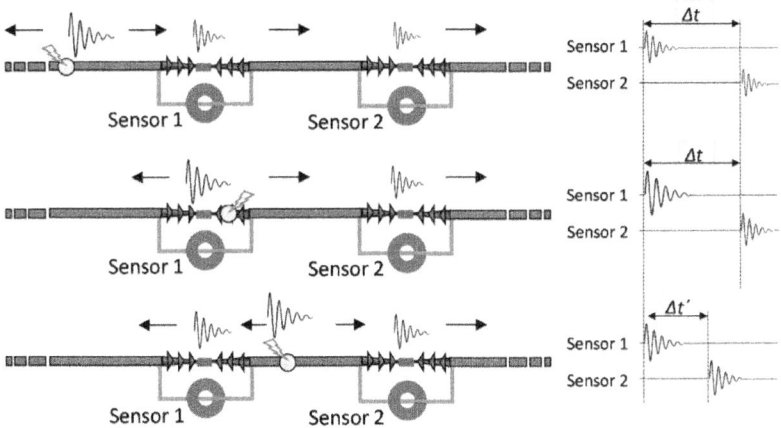

Figure 25: (a) Defect before the section of cable between sensors; (b) Defect located in the same emplacement as one of the sensors; (c) Defect inside the section of cable between sensors.

All clusters obtained with the classification tool by pulse waveform have been analyzed except cluster 6, see Figure 20a and Figure 26a. The PD source corresponding with this cluster was detected with the sensor HFCT 1 but not with the HFCT 2 due to the attenuation effect of the pulses, so the location of this source was not possible. Analyzing the characteristics of the pattern it seems that it is related with a corona defect (tip as ground electrode), probably located in the GIS compartment.

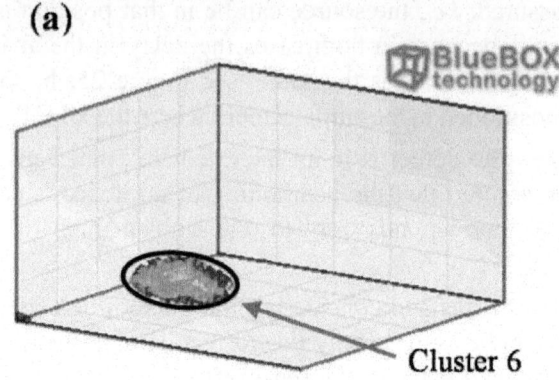

Cluster 6

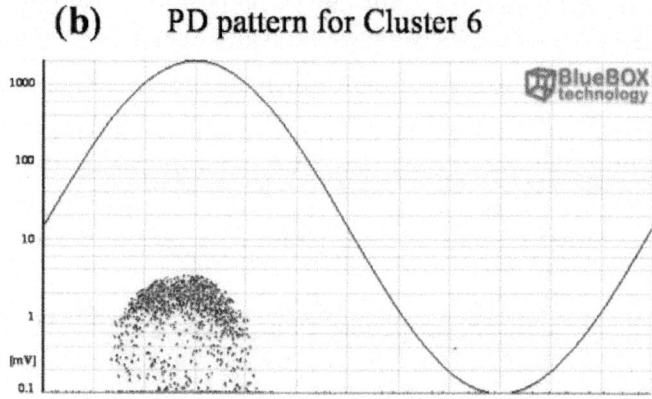

Figure 26: (a) Cluster 6 detected with the sensor HFCT 1; (b) PRPD pattern corresponding to cluster 6.

Once analyzed the results obtained with the measurements performed applying the HFCT sensors the following statements can be made.

- In position C two types of sources were identified: corona in air and an internal defect.

- The location of the corona defect in position C was ratified. However, the emplacement of the internal defect could not be confirmed.

- In position D an internal defect in a joint was detected and located.

- A corona type PD source related with a tip referenced to ground (cluster 6) was detected with the sensor HFCT 1. The location of this source could not be determined.

In order to complement the previous diagnosis, the measurements performed with the invasive UHF sensors were analyzed. Furthermore, to confirm the location of the internal defect in position C, the mobile non-invasive sensor UHF 3 was coupled in each cable termination of this emplacement.

The PRPD patterns for the acquisition with the invasive couplers located in positions A and B of the GIS compartment are shown in Figure 27a,b. Due to the selective approach of the measurements in UHF only a single pattern has been obtained for each sensor. Moreover, it can be noticed the immunity that this technique presents to background noise interferences.

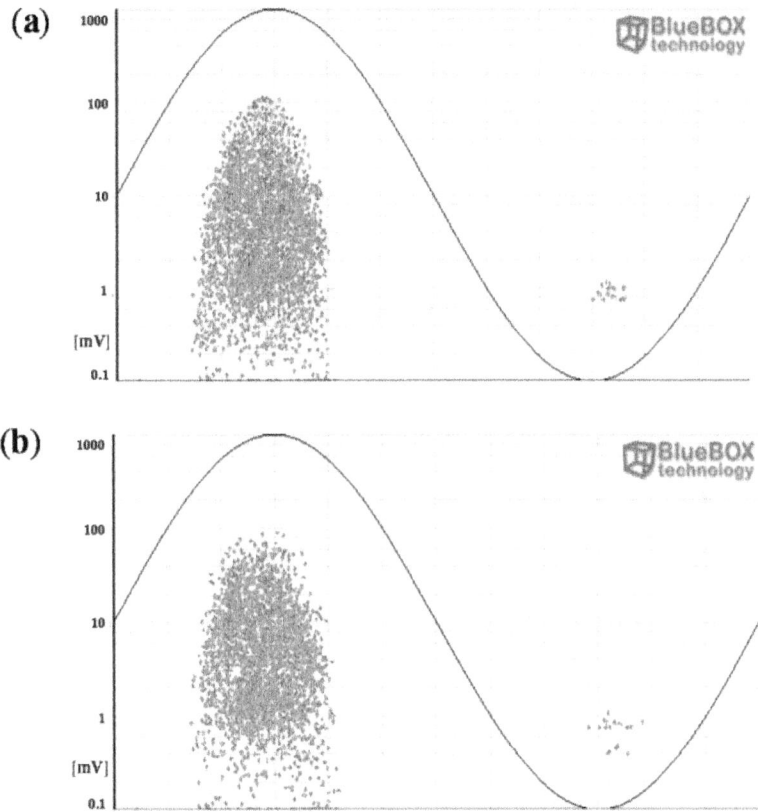

Figure 27: PRPD patterns obtained with the UHF sensors. (a) Sensor UHF 1 in position A; (b) Sensor UHF 2 in position B.

The PRPD pattern detected with the invasive UHF sensors in the GIS compartment is characteristic of a corona defect in SF_6 with a tip referenced to ground. PD pulses appear on the crests of the reference voltage signal and there are more pulses in the positive half-cycle.

As the captures were performed with invasive UHF sensors, it can be assured that the pulses were generated inside the compartment; UHF sensors are insensitive to the low frequency spectrum of the pulses generated in the cable system that propagate to the GIS. Furthermore, this PRPD pattern has the same shape as the one measured with the sensor HFCT 1 corresponding with cluster 6, so it can be concluded that the non-localized corona defect measured with this sensor is the one generated in the GIS chamber.

Due to the high propagation speed of PD signals inside the GIS compartment (0.28 m/ns) and the short distance between sensors (2 m), in this case the localization tool can not be applied, as the sampling period of the measuring instrument is 1 sample every 10ns. An approximate location of the defect is possible comparing the amplitude of the integrated pulse of a PD measured with both sensors; the coupler with the higher signal level is closer to the PD source. The higher amplitude obtained for the pulse measured with the sensor UHF 1 evidences that the defect is closer to the position A, see Figure 28. An accurate location can be obtained by analyzing the travelling time of the UHF signals measured with a dual-channel oscilloscope of at least 1 GHz of bandwidth and 5 GS/s of sampling rate.

Finally, the PRPD pattern shown in Figure 29 was obtained with the mobile sensor UHF 3 coupled in the right cable termination of position C. No PD signals were detected in the left termination. This pattern is characteristic of an internal defect and is very similar to the one obtained with the sensor HFCT 1, see Figure 23c. The detection with the sensor UHF 3 of this internal pattern confirms that the PD source is in this cable termination.

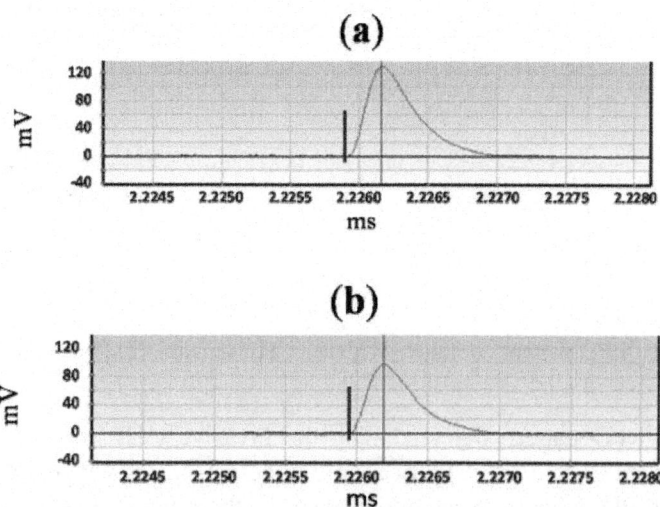

Figure 28: (a) PD pulse measured with the sensor UHF 1 and converted into a HF pulse; (b) PD pulse measured with the sensor UHF 2 and converted into a HF pulse.

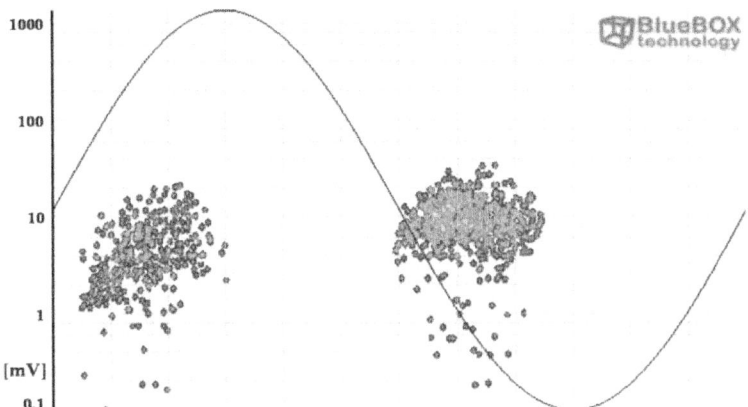

Figure 29: PRPD pattern obtained with the mobile non-invasive UHF sensor coupled in the right termination of position C.

The measurements performed with the UHF sensors made it possible to identify and locate correctly the corona defect of the GIS compartment and also to confirm the emplacement of the internal defect in the right cable termination of position C. The complementary measurements with both techniques enabled the identification and location of all the insulation defects generated in the HV installation.

CONCLUSIONS

In the study presented theoretical and practical aspects about the application of HF and UHF sensors for PD measurements have been considered and analyzed. The use of HFCT sensors with a bandwidth of 20 MHz is recommended for their practical implementation in on-line PD measurements.

For the measurements performed with UHF sensors the use of a UHF-HF converter enables to take advantages of capturing the PD pulses in the UHF range and to process them in the HF range. Therefore, using only a PD instrument with a bandwidth covering the HF range, the signals measured with HF and UHF sensors can be analyzed. Several insulation defects present simultaneously in a HV installation can be detected and located applying the measuring and analysis procedure described in this paper.

Representative measurements have been performed with the PD instrument developed, in order to show the effectiveness and advantages offered by the complementary measurements with HF and UHF sensors. The information provided measuring in the HF and UHF frequency ranges and the analysis of all the individual PRPD patterns obtained provide the knowledge necessary to make accurate diagnoses.

The measurement and analysis techniques presented, with the combination of the different sensors and processing tools can be employed to accurately assess the insulation condition of HV grid network assets in on-line measurements or in continuous monitoring applications.

ACKNOWLEDGMENTS

The authors would like to thank DIAEL Corporation for their continuous support in the development of the measuring techniques which have been presented and are integrated as part of the BlueBOX technology.

AUTHOR CONTRIBUTIONS

The presented study was developed by the authors contributing each one with a specific task. Fernando Álvarez carried out the analysis of the technical details and practical implementations of HFCT and UHF sensors for their application in PD measurements. He also performed the experimental works and evaluated the results obtained. Fernando Garnacho focused the interest of the paper and supported Fernando Álvarez in the aforementioned activities. Javier Ortego and Miguel Ángel Sanchez-Urán contributed in the development of the measuring instrument and application software for the PD measurement and processing.

REFERENCES

1.   Wester, F. Condition Assessment of Power Cables Using PD Diagnosis at Damped AC Voltages. Ph. D. Thesis, Delft University of Technology, Delft, The Netherlands, 2004; pp. 11–35.

2.   Meijer, S.; Gulski, E.; Smit, J.J. Pattern Analysis of Partial Discharges in SF_6 GIS. *IEEE Trans. Dielectr. Electr. Insul.* 1998, 5, 830–842.

3.   Fuhr, J. Procedure for Identification and Localization of Dangerous PD Sources in Power Transformers. *IEEE Trans. Dielectr. Electr. Insul.* 2005, 12, 1005–1014.

4.   Haddad, A.; Warne, D.F. *Advances in High Voltage Engineering*, 2nd ed.; IET Power and Energy Series; IET: London, UK, 2007; pp. 37–190.

5.   Tian, Y.; Lewin, P.; Davies, A. Comparison of On-line Partial Discharge Detection Methods for HV Cable Joints. *IEEE Trans. Dielectr. Electr. Insul.* 2002, 9, 604–615.

6.   Wang, X.; Li, B.; Roman, H.; Russo, O.L.; Chin, K.; Farmer, K.R.

Acousto-optical PD Detection for Transformers. *IEEE Trans. Power Deliv.* 2006, *21*, 1068–1073.

7.   Wang, L.; Fang, N.; Wu, C.; Qin, H.; Huang, Z. A Fiber Optic PD Sensor Using a Balanced Sagnac Interferometer and an EDFA-Based DOP Tunable Fiber Ring Laser. *Sensors* 2014, *14*, 8398–8422.

8.   Posada-Roman, J.; Garcia-Souto, J.A.; Rubio-Serrano, J. Fiber Optic Sensor for Acoustic Detection of Partial Discharges in Oil-Paper Insulated Electrical Systems. *Sensors* 2012, *12*, 4793–4802.

9.   Rodrigo, A.; Llovera, P.; Fuster, V.; Quijano, A. Influence of High Frequency Current Transformers Bandwidth on Charge Evaluation in Partial Discharge Measurements. *IEEE Trans. Dielectr. Electr. Insul.* 2011, *18*, 1798–1802.

10. Rodrigo, A.; Llovera, P.; Fuster, V.; Quijano, A. Study of Partial Discharge Charge Evaluation and the Associated Uncertainty by Means of High Frequency Current Transformers. *IEEE Trans. Dielectr. Electr. Insul.* 2012, *19*, 434–442.

11. Stone, G. Partial Discharge—Part VII. Practical Techniques for Measuring PD in Operating Equipment. *IEEE Electr. Insul. Mag.* 1991, *7*, 9–19.

12. Cavallini, A.; Montanari, G.; Contin, A.; Pulletti, F. A New Approach to the Diagnosis of Solid Insulation Systems Based on PD Signal Inference. *IEEE Electr. Insul. Mag.* 2003, *19*, 23–30.

13. Ardila-Rey, J.; Martínez-Tarifa, J.M.; Robles, G.; Rojas-Moreno, M. Partial Discharge and Noise Separation by Means of Spectral-Power Clustering Techniques. *IEEE Trans. Dielectr. Electr. Insul.* 2013, *20*, 1436–1443.

14. Ardila-Rey, J.A.; Rojas-Moreno, M.V.; Martínez-Tarifa, J.M.; Robles, G. Inductive Sensor Performance in Partial Discharges and Noise Separation by Means of Spectral Power Ratios. *Sensors* 2014, *14*, 3408–3427.

15. Gulski, E.; Cichecki, P.; Wester, F.; Smit, J.; Bodega, R.; Hermans, T.; Seitz, P.B.; Quak, B.; Vries, F. On-Site Testing and PD Diagnosis of High Voltage Power Cables. *IEEE Trans. Dielectr. Electr. Insul.* 2008, *15*, 1691–1700.

16. Kreuger, F.H.; Gulski, E.; Krivda, A. Classification of Partial Discharges. *IEEE Trans. Dielectr. Electr. Insul.* 1993, *28*, 917–931.

17. International Standard IEC 60270. In *High Voltage Test Techniques—Partial Discharge Measurements*, 3rd ed.; International Electrotechnical Commission: Geneva, Switzerland, 2000.

18. Lemke, E. *Guide for Partial Discharge Measurement in Compliance to*

IEC 60270 Std, CIGRE Technical Brochure; CIGRE, 2008. Available online: http://www.e-cigre.org/Order/select.asp?ID=13723 (accessed on 23 March 2015).

19. Tumanski, S. Induction coil sensors: A review. *Meas. Sci. Technol.* 2007, *18*, 31–46.

20. IEEE Std. 400.3. In *IEEE Guide for Partial Discharge Testing of Shielded Power Cable Systems in a Field Environment*; IEEE: New York, NY, USA, 2006.

21. Shibuya, Y.; Matsumoto, S.; Tanaka, M.; Muto, H.; Kaneda, Y. Electromagnetic Waves from Partial Discharges and their Detection Using Patch Antenna. *IEEE Trans. Dielectr. Electr. Insul.* 2010, *17*, 862–871.

22. Reid, A.J.; Judd, M.D.; Stewart, B.G.; Fouracre, R.A. Partial discharge current pulses in SF_6 and the effect of superposition of their radiometric measurement. *J. Phys. D: Appl. Phys.* 2006, *39*, 4167–4177.

23. Lemke, E.; Gulski, E.; Hauschild, W.; Malewski, R.; Mohaupt, P.; Muhr, M.; Rickmann, J.; Strehl, T.; Wester, F.J.*Practical Aspects of the Detection and Location of Partial Discharges in Power Cables*; CIGRE Technical Brochure. CIGRE, 2005. Available online: http://www.e-cigre. org/Order/select.asp?ID=9303 (accessed on 23 March 2015).

24. Judd, M.D.; Yang, L.; Hunter, I.B.B. Partial Discharge Monitoring of Power Transformers Using UHF Sensors. Part I: Sensors and Signal Interpretation. *IEEE Electr. Insul. Mag.* 2005, *21*, 5–14.

25. Lopez-Roldan, J.; Tang, T.; Gaskin, M. Optimisation of a Sensor for Onsite Detection of Partial Discharges in Power Transformers by the UHF Method. *IEEE Trans. Dielectr. Electr. Insul.* 2008, *15*, 1634–1639.

26. Hoshino, T.; Kato, K.; Hayakawa, N.; Okubo, H. A Novel Technique for Detecting Electromagnetic Wave Caused by Partial Discharge in GIS. *IEEE Trans. Power Deliv.* 2001, *16*, 545–551.

27. Kaneko, S.; Okabe, S.; Yoshimura, M.; Muto, H.; Nishida, C.; Kamei, M. Detecting Characteristics of Various Type Antennas on Partial Discharge Electromagnetic Wave Radiating Through Insulating Spacer in Gas Insulated Switchgear. *IEEE Trans. Dielectr. Electr. Insul.* 2009, *16*, 1462–1472.

28. Raja, K.; Devaux, F.; Lelaidier, S. Recognition of Discharge Sources Using UHF PD Signatures. *IEEE Electr. Insul. Mag.*2002, *18*, 8–14.

29. Judd, M.D.; Farish, O.; Hampton, B.F. Broadband couplers for UHF detection of partial discharge in gas-insulated substations. *IEE Proc. Sci.*

Meas. Technol. 1995, *142*, 237–243.

30. Li, T.; Rong, M.; Zheng, C.; Wang, X. Development Simulation and Experiment Study on UHF Partial Discharge Sensor in GIS. *IEEE Trans. Dielectr. Electr. Insul.* 2012, *19*, 1421–1430.

31. Hikita, M.; Ohtsuka, S.; Teshima, T.; Okabe, S.; Kaneko, S. Electromagnetic (EM) Wave Characteristics in GIS and Measuring the EM Wave Leakage at the Spacer Aperture for Partial Discharge Diagnosis. *IEEE Trans. Dielectr. Electr. Insul.* 2007, *14*, 453–460.

32. Watkins, J. Circular resonant structures in microstrip. *Electron. Lett.* 1969, *5*, 524–525.

33. Shen, L.; Long, S.A.; Allerding, M.; Walton, M. Resonant Frequency of a Circular Disc, Printed-Circuit Antenna. *IEEE Trans. Antennas Propag.* 1977, *25*, 595–596.

34. Long, S.A.; Shen, L.C.; Morel, P.B. Theory of the circular-disc printed-circuit antenna. *Proc. Inst. Electr. Eng.* 1978, *125*, 925–928.

35. Zhang, S.; Zheng, X.; Zhang, J.; Cao, H.; Zhang, X. Study of GIS Partial Discharge On-line Monitoring Using UHF Method. In Proceedings of the International Conference on Electrical and Control Engineering (ICECE), Wuhan, China, 25–27 June 2010; pp. 4262–4265.

36. Shu, E.W.; Boggs, S. Dispersion and PD Detection in Shielded Power Cable. *IEEE Electr. Insul. Mag.* 2008, *24*, 25–29.

37. Garnacho, F.; Sánchez-Urán, M.A.; Ortego, J.; Álvarez, F.; Perpiñán, O.; Puelles, E.; Moreno, R.; Prieto, D.; Ramos, D. New Procedure to Determine Insulation Condition of High Voltage Equipment by Means of PD Measurements in Service. In Proceedings of 44th International CIGRE Session, Paris, France, 26–31 August 2012.

38. Shim, I.; Soraghan, J.J.; Siew, W.H. Detection of PD Utilizing Digital Signal Processing Methods. Part 3: Open-Loop Noise Reduction. *IEEE Electr. Insul. Mag.* 2001, *17*, 6–13.

39. Ma, X.; Zhou, C.; Kemp, I.J. Automated Wavelet Selection and Thresholding for PD Detection. *IEEE Electr. Insul. Mag.* 2002, *18*, 37–45.

40. Meijer, S. Partial Discharge Diagnosis of High-Voltage Gas-Insulated Systems. Ph.D. Thesis, Delft University, Delft, The Netherlands, November 2001.

Chapter 9

CALCULATION OF CONSTITUTIVE PARAMETERS FROM ELECTRIC AND MAGNETIC FIELD MEASUREMENTS IN AN ANISOTROPIC MEDIUM WITH A TRIAXIAL INSTRUMENT

Ertan Pekşen

Department of Geophysical Engineering, Kocaeli University, Kocaeli, Turkey

ABSTRACT

A hypothetical electric and magnetic induction tensor is considered in an anisotropic medium. The sources are magnetic dipoles. In such a medium, constitute parameters can be calculated by combining electric and magnetic field measurements. Constitutive parameters are not a scalar in this case. They are tensors, so parameters have at least both horizontal and vertical components in a uniaxial medium. These calculated parameters from the field measurement are horizontal and vertical conductivity, permittivity, and magnetic permeability. Operating frequency range is also quite large. It is up to 4 GHz. A hypothetical instrument should measure gradient fields both electric and magnetic types as well.

INTRODUCTION

Conductivity, permittivity, and permeability of a medium are known constitutive parameters [1]. These parameters are used to interpret the subsurface with different geophysical methods and well-logging types. Relative magnitudes of these parameters are quite distinct as very well-known properties of rocks [2-5]. Since these parameters seldom vary for most rocks and minerals, most scientists generally assume dielectric permittivity and magnetic permeability of the medium as free space for convenience. In an isotropic medium, constitutive parameters are the same in all directions; thus, they are scalars in that case. However, in an anisotropic medium these parameters depend on their directions, so they must at least be a vector. In the present study, an anisotropic medium is considered; thus, the determination of this kind medium requires five parameters in a vertical well. These are horizontal and vertical conductivities, permittivities, and magnetic permeability. This kind of anisotropy is known as a uniaxial anisotropy. In a Cartesian coordinate system, constitute parameters are the same in the x and y directions, but they are different in the z direction. If conductivity values are different in the x, y and z directions, this sort of medium is called as a biaxial medium.

In induction well logging, the electric anisotropy affects field measurements. Klein et al. [6] realized this effect and they stated that the possible reason for high reading could be the electrical anisotropy. Obviously, the effects of electrical anisotropy make interpretation erroneous. To deal with this effect and eliminate from the data, the electric anisotropy is considered.

In a wellbore, the field is also affected from a relative deviation angle. These effects can be included/extracted to/from the fields by rotating the Euler's rotation matrix. Zhdanov et al. [7] studied the tensor well logging and they generalized Doll's idea for an anisotropic medium. They showed that from the field components one can calculate not only conductivity values (horizontal and vertical) but also relative deviation and bearing angles. They used imaginary components of magnetic fields. Zhang et al. [8] showed that using field measurements one can determine relative angles such as relative dip and relative bearing as well. In general, there are three independent coordinate systems: a well, earth, and instrument coordinate systems. In this study, I consider a whole space (a uniaxial medium) and instrument axis are coincided with each other's for the simplicity.

Many scientist [9-13] studied uniaxial media due to its simplicity. There have been a few authors studied a biaxial anisotropic medium, which can be characterized by its conductivity in each different directions in a Cartesian coordinate system [14-16].

A very few authors have studied electric field, and electric and magnetic field measurement together. Gribenko and Zhdanov [17] considered combination of electric and magnetic fields. They stated that the combination of the electric and magnetic field measurement gives better results with respect to conductivity distribution of a well. Here, the idea of combination of electric and magnetic field can be extended to an induction logging problem in an anisotropic medium.

In this study, I combine electric and magnetic fields in an anisotropic whole space. This is a hypothetical instrument. I demonstrate that constitutive parameters can be calculated by combining electric and magnetic field measurement in an anisotropic medium. Parameters are vertical and horizontal conductivities, dielectric permittivities, and magnetic permeability. These parameters characterize the field behaviors in a corresponding medium. In general, our aim is to measure field components and estimate those parameters, which characterize the field behaviors. In the present study, neither relative dip nor bearing is considered in the medium. In general, this hypothetical instrument should measure full electric and magnetic field components and their gradient in a well. Zhdanov [18] gave a definition of a gradient type measurement in an anisotropic medium for magnetic field. The same idea can also be extended to electric field measurement as well.

MAXWELL'S EQUATIONS IN AN ANISOTROPIC MEDIUM

Moran and Gianzero [11] studied induction well logging in a uniaxial anisotropic medium. Maxwell's equations are given in such a medium as:

$$\nabla \times \mathbf{H} = \hat{\sigma} \cdot \mathbf{E},\tag{1}$$

$$\nabla \times \mathbf{E} = i\omega\mu_0\mathbf{H} + i\omega\mu_0\mathbf{M},\tag{2}$$

where E is the electric field (V/m), H is the magnetic field (A/m). M is a magnetic dipole moment (Am^2). The conductivity tensor in a transversely isotropic (TI) medium is:

$$\hat{\sigma} = \begin{bmatrix} \sigma_h - i\omega\varepsilon_h & 0 & 0 \\ 0 & \sigma_h - i\omega\varepsilon_h & 0 \\ 0 & 0 & \sigma_v - i\omega\varepsilon_v \end{bmatrix},\tag{3}$$

where σ_h is the horizontal component and σ_v is the vertical component of conductivity tensor. ε_h and ε_v are the horizontal and vertical permittivity of the medium, respectively. The free space dielectric permittivity is $\varepsilon_0 = 8.854 \times 10^{-12}$ F/m. The free space magnetic permeability is $\mu_0 = 4\pi \times 10^{-7}$ H/m. The transverse isotropic medium is excited through an electromagnetic field generated by magnetic dipoles with unit moments. The time dependence is $e^{-i\omega t}$. f is the frequency of a source in Hertz.

Moran and Gianzero [11] solved Equations (1) and (2) by using the Hertz potential. In this paper, I assume that the dielectric permittivity and magnetic permeability of the medium are not free space. In this situation, electric and magnetic field components can be written:

$$\mathbf{E}_m^n = i\omega\mu\left(\sigma_h - i\omega\varepsilon_h\right)\hat{\sigma}^{-1} \cdot \nabla \times \boldsymbol{\pi},\tag{4}$$

$$\mathbf{H}_m^n = i\omega\mu\hat{\sigma} \cdot \boldsymbol{\pi} + \nabla\Phi,\tag{5}$$

where π is a Hertz vector and Φ is a scalar potential. The medium dielectric permittivity and magnetic permeability are $\varepsilon_h = \varepsilon_h^r \varepsilon_0$, $\varepsilon_v = \varepsilon_v^r \varepsilon_0$ and $\mu = \mu_r \mu_0$, respectively. ε_h^r and ε_v^r stand for the relative dielectric permittivity, μ_r represents a relative magnetic permeability. Super and subscripts depict a magnetic moment and receiver direction along the corresponding axis; thus, n and m can be x, y, and z, in Equations (4) and (5).

In this paper, I use as the following notation: Bold face symbols or characters with a hat stand for tensors, while bold face symbols or characters without a hat illustrate vectors. Any characters with italic are a scalar. Therefore, calculations involve some dyad algebra such as dyadicvector dot product in Equations (4) and (5). Next two sections investigate electric and magnetic field components.

ELECTRIC INDUCTION TENSOR

Electric induction tensor has 9 components in a general case. Three orthogonal receivers and transmitters are elongated with their corresponding axis in a Cartesian coordinate system. The sources are magnetic dipoles and receivers are electromagnetic sensor or coils. To derive the electric field components is a straightforward calculation that will not be repeated here. Reader can find all electric components from [19-21].

The electric induction tensor can be represented in a matrix form as:

$$\hat{E} = \begin{bmatrix} E_x^x & E_x^y & E_x^z \\ E_y^x & E_y^y & E_y^z \\ E_z^x & E_z^y & E_z^z \end{bmatrix}, \tag{6}$$

where superscripts are magnetic dipole orientations and subscripts are receiver orientations. Bear in mind that the sources are magnetic dipoles. Figure 1 displays a hypothetical electric and magnetic induction tensor. Figure 2 shows the behavior of electric field components in a x-y plane. For field simulation, the horizontal and vertical conductivities are 0.1 and 0.025 S/m, respectively. The operating frequency is 20 kHz. The field computation is conducted with free space dielectric permeability and magnetic permeability. M is a magnetic dipole with a unit. One can realize that the symmetrical fields have the same amplitude, but different sings.

MAGNETIC INDUCTION TENSOR

In this section, a magnetic induction tensor is investigated. As the electric induction tensor, the magnetic induction tensor has also 9 components.

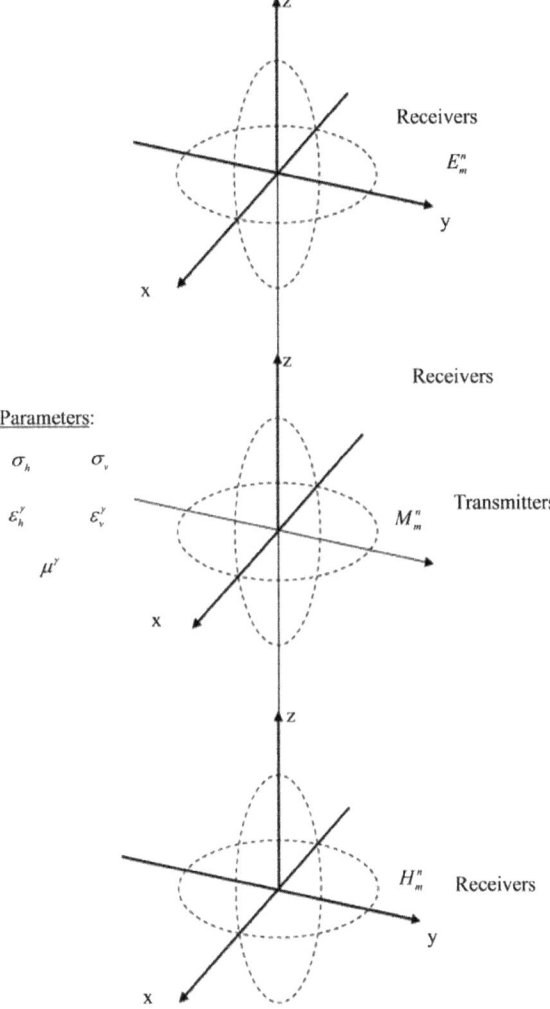

Figure 1: A sketch of a triaxial induction logging instrument is in a well. The physical medium properties are conductivity, dielectric permittivity, and magnetic permeability. The sources are magnetic dipoles. Receives are both magnetic and electric field sensors. Super and subscripts indicate the direction of transmitter and receiver, respectively (n and m can be x, y and z). Conductivity and dielectric permittivity are tensors, since they have principal values at least two values. The magnetic permeability is assumed a scalar.

It can be shown in a matrix form as:

$$\hat{\mathbf{H}} = \begin{bmatrix} H_x^x & H_x^y & H_x^z \\ H_y^x & H_y^y & H_y^z \\ H_z^x & H_z^y & H_z^z \end{bmatrix},$$

(7)

where the notation is the same as the electric induction tensor. Three transmitter and receiver coils are oriented along the x, y, and z-axes in a Cartesian coordinate system. Derivation of magnetic field components is a straightforward calculation that will not be repeated here. Magnetic field of analytic components can be found from [7,22].

The magnetic induction tensor is symmetric. Behavior of magnetic field components is displayed inFigure 3. The uniaxial model parameters are the same as the previous electric field calculation. Thanks to commercially available 3DEX and Rt Scanner induction well logging instruments reads a lot of data at different frequencies for various field components. That means considerable amount of data are available. From 3DEX and Rt Scanner data, conductivities, relative dip and bearing angles can be estimated [8].

In Figure 4, each components of magnetic induction tensor are displayed in volumetric 3-D at 20 kHz. In all panels, imaginary parts of the magnetic field are normalized by absolute value of the corresponding components. The medium conductivities are: $\sigma_h = 1/40$ S/m and $\sigma_h = 1/80$ S/m. To able to see all panels, 0.1 value used for off diagonal components. As for diagonal components, I use 0.0045 value. Red and blue colors show sign. Blue used for negative, while red used for positive values. The behaviors of components are quite different from each other's.

ESTIMATION OF EARTH PARAMETERS: CONDUCTIVITY, DIELECTRIC PERMITTIVITY, AND MAGNETIC PERMEABILITY

I consider a hypothetical multi-component induction well logging instrument. As mentioned previously, the instrument has magnetic dipoles as a source in the x, y and z directions. Receivers should measure magnetic and electric field components. It is a combination of electric and magnetic induction tensor components. It requires some tensor-gradient measurements. From this hypothetical electric and magnetic induction tensor-gradient measurement, one can estimate earth (or formation) parameters. There are many different formulas are given in Table 1. In the first column of the table is a magnetic moment direction, while the first row depicts earth parameters. A simple derivation is given in Appendix A. From Table 1, it is easy to see that there are many different formulas for estimating the earth parameters. One can calculate conductivities and dielectric permitivities in the horizontal and vertical directions. These formulas can be useful when one of these components is very noise, the other one can be used for calculating the corresponding earth parameters in a practical situation. They may be useful for checking parameters against each other using different field measurements. This method will obviously increase the quality of the formation evaluation. Table 1 has formulas for conductivities and dielectric permittivities. Appendix B derives formulas for magnetic permeability. These formulas are given in Table 2. In both tables, $\Re$ and $\Im$ stand for real and imaginary components of the corresponding field, respectively.

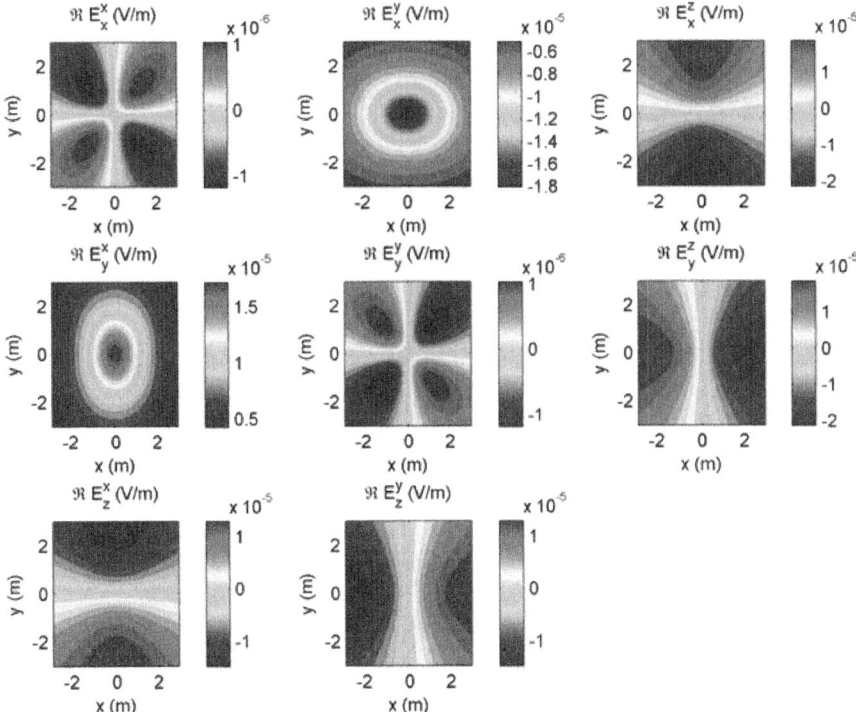

Figure 2: Real parts of the magnetic field components are calculated in an anisotropic medium with 20 kHz operating frequencies at $z = 1$ m. The horizontal and vertical conductivities are 1/40 and 1/80 S/m, respectively. It is a top view of a whole space.

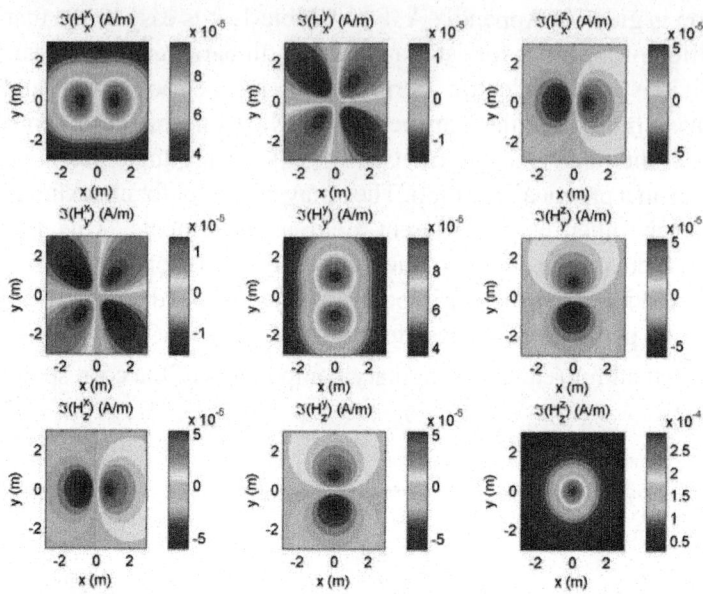

Figure 3: Imaginary parts of the magnetic field components are calculated in an aniso-tropic medium with 20 kHz operating frequencies at z = 1 m. The horizontal and verti-cal conductivities are 1/40 and 1/80 S/m, respectively. It is a top view of the medium.

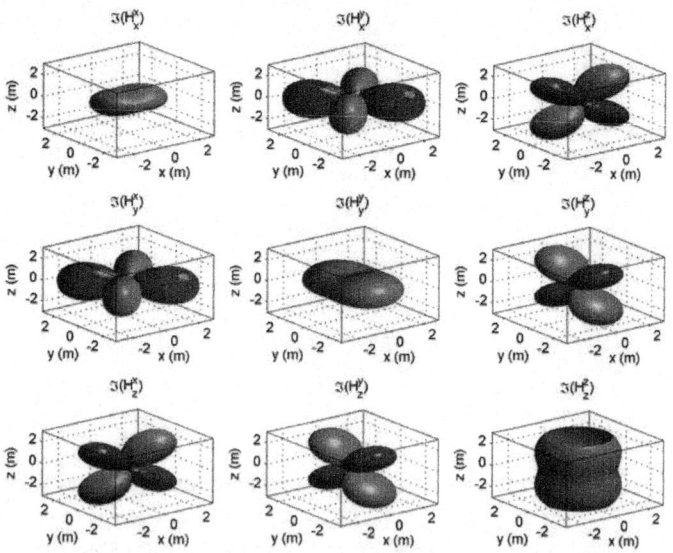

Figure 4: Volumetric 3-D representation of the magnetic tensor components at value 0.1 (off diagonal) and 0.0045 (diagonal). The medium is conducted with σ_h = 1/40 S/m and σ_h = 1/80 S/m. Blue colors display negative, while red ones depict positive values of imaginary part of the magnetic field.

Table 1: Magnetic moments are in the first column. The superscripts show the direction of the magnetic moments. The rest of the formulas are apparent conductivities and dielectric permittivities. $\Re$ and $\Im$ depict for real and imaginary components of the corresponding field

		σ_u		σ_v	ε_u		ε_v
M^x	$\Re\left\{\dfrac{\left(\dfrac{\partial H_y^r}{\partial y}-\dfrac{\partial H_z^r}{\partial z}\right)}{E_x^r}\right\}$	$\Re\left\{\dfrac{\left(\dfrac{\partial H_y^r}{\partial x}-\dfrac{\partial H_x^r}{\partial z}\right)}{E_y^r}\right\}$	$\Re\left\{\dfrac{\left(\dfrac{\partial H_y^r}{\partial x}-\dfrac{\partial H_x^r}{\partial y}\right)}{E_z^r}\right\}$	$\Im\left\{\dfrac{\left(\dfrac{\partial H_z^i}{\partial y}-\dfrac{\partial H_y^i}{\partial z}\right)}{i\omega E_x^i}\right\}$	$\Im\left\{\dfrac{\left(\dfrac{\partial H_z^i}{\partial x}-\dfrac{\partial H_x^i}{\partial z}\right)}{i\omega E_y^i}\right\}$	$\Im\left\{\dfrac{\left(\dfrac{\partial H_y^i}{\partial x}-\dfrac{\partial H_x^i}{\partial y}\right)}{i\omega E_z^i}\right\}$
M^y	$\Re\left\{\dfrac{\left(\dfrac{\partial H_z^r}{\partial y}-\dfrac{\partial H_y^r}{\partial z}\right)}{E_x^r}\right\}$	$\Re\left\{\dfrac{\left(\dfrac{\partial H_z^r}{\partial x}-\dfrac{\partial H_x^r}{\partial z}\right)}{E_y^r}\right\}$	$\Re\left\{\dfrac{\left(\dfrac{\partial H_y^r}{\partial x}-\dfrac{\partial H_x^r}{\partial y}\right)}{E_z^r}\right\}$	$\Im\left\{\dfrac{\left(\dfrac{\partial H_z^i}{\partial y}-\dfrac{\partial H_y^i}{\partial z}\right)}{i\omega E_x^i}\right\}$	$\Im\left\{\dfrac{\left(\dfrac{\partial H_z^i}{\partial x}-\dfrac{\partial H_x^i}{\partial z}\right)}{i\omega E_y^i}\right\}$	$\Im\left\{\dfrac{\left(\dfrac{\partial H_y^i}{\partial x}-\dfrac{\partial H_x^i}{\partial y}\right)}{i\omega E_z^i}\right\}$
M^z	$\Re\left\{\dfrac{\left(\dfrac{\partial H_z^r}{\partial y}-\dfrac{\partial H_y^r}{\partial z}\right)}{E_x^r}\right\}$	$\Re\left\{\dfrac{\left(\dfrac{\partial H_z^r}{\partial x}-\dfrac{\partial H_x^r}{\partial z}\right)}{E_y^r}\right\}$	$\Re\left\{\dfrac{\left(\dfrac{\partial H_y^r}{\partial x}-\dfrac{\partial H_x^r}{\partial y}\right)}{E_z^r}\right\}$	$\Im\left\{\dfrac{\left(\dfrac{\partial H_z^i}{\partial y}-\dfrac{\partial H_y^i}{\partial z}\right)}{i\omega E_x^i}\right\}$	$\Im\left\{\dfrac{\left(\dfrac{\partial H_z^i}{\partial x}-\dfrac{\partial H_x^i}{\partial z}\right)}{i\omega E_y^i}\right\}$	$\Im\left\{\dfrac{\left(\dfrac{\partial H_y^i}{\partial x}-\dfrac{\partial H_x^i}{\partial y}\right)}{i\omega E_z^i}\right\}$

Numerical Example

In a uniaxial medium, I consider as the following parameters for numerical calculation: $\sigma_u = 0.1$ and $\sigma_v = 0.025$ S/m; $\varepsilon_r^h = 27$ and $\varepsilon_r^v = 7$; $\mu_r = 255$. The frequency range for calculation begins at 10 kHz and it expands up to 4 GHz. The distance between transmitter and receivers is 1.6 m as a typical induction T-R distance.

The model response in a whole space is calculated by a magnetic dipole oriented in the y direction. Electric and magnetic field components and their space derivatives are calculated by using analytic formulas. For constitute parameter calculation, a group of the formulas given in **Table 1** are used, which are on the second rows. All parameters are related to a uniaxial medium estimated, correctly. Figure 5 displays the result. Note that these are only some simple formulas in Tables 1 and 2.

Table 2: Magnetic moments are in the first column. The superscripts show the direction of the magnetic moments. All formulas in the table can be used for calculating apparent magnetic permeability

		μ	
M^x	$\dfrac{1}{i\omega}\dfrac{\left(\dfrac{\partial E_z^r}{\partial y}-\dfrac{\partial E_y^r}{\partial z}\right)}{M^x+H_x^r}$	$\dfrac{1}{i\omega}\dfrac{\left(\dfrac{\partial E_x^r}{\partial x}-\dfrac{\partial E_x^r}{\partial z}\right)}{M^x+H_y^r}$	$\dfrac{1}{i\omega}\dfrac{\left(\dfrac{\partial E_y^r}{\partial x}-\dfrac{\partial E_x^r}{\partial y}\right)}{M^x+H_z^r}$
M^y	$\dfrac{1}{i\omega}\dfrac{\left(\dfrac{\partial E_z^r}{\partial y}-\dfrac{\partial E_y^r}{\partial z}\right)}{M^y+H_x^r}$	$\dfrac{1}{i\omega}\dfrac{\left(\dfrac{\partial E_x^r}{\partial x}-\dfrac{\partial E_z^r}{\partial z}\right)}{M^y+H_y^r}$	$\dfrac{1}{i\omega}\dfrac{\left(\dfrac{\partial E_y^r}{\partial x}-\dfrac{\partial E_x^r}{\partial y}\right)}{M^y+H_z^r}$
M^z	$\dfrac{1}{i\omega}\dfrac{\left(\dfrac{\partial E_z^r}{\partial y}-\dfrac{\partial E_y^r}{\partial z}\right)}{M^z+H_x^r}$	$\dfrac{1}{i\omega}\dfrac{\left(\dfrac{\partial E_x^r}{\partial x}-\dfrac{\partial E_z^r}{\partial z}\right)}{M^z+H_y^r}$	$\dfrac{1}{i\omega}\dfrac{\left(\dfrac{\partial E_y^r}{\partial x}-\dfrac{\partial E_x^r}{\partial y}\right)}{M^z+H_z^r}$

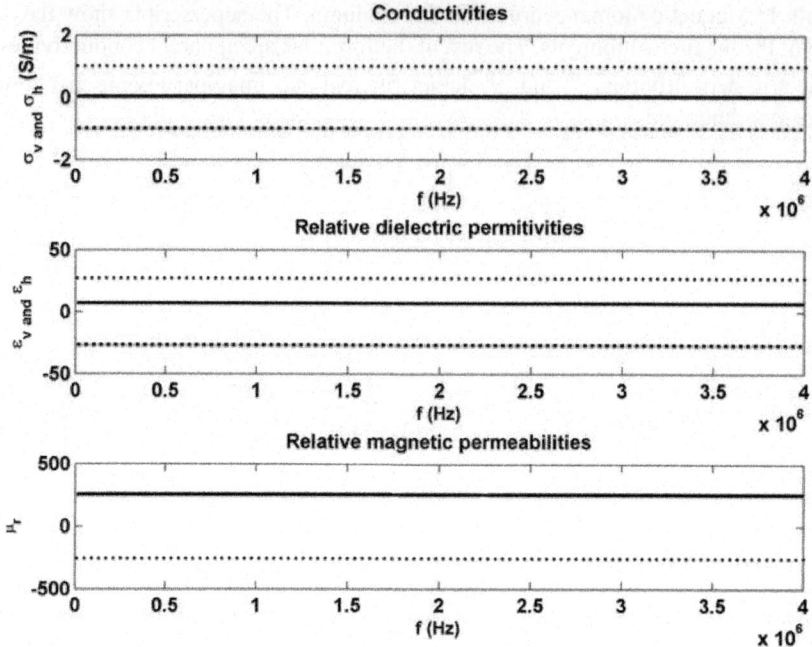

Figure 5: Constitutive parameters are horizontal and vertical conductivity, relative dielectric permittivity and magnetic permeability of the medium. The magnetic moment direction is in the y direction. On each panel, solid, dotted, and dashed lines show the result. The calculation done with the formulas are given on the second row in Table 1. Solid lines illustrate horizontal values of both conductivity and dielectric, while dotted and dashed lines display vertical values of those parameters. The frequency range is quite large between at 10 kHz and 4 GHz. The medium has conductivity values with $\sigma_h = 0.1$ and $\sigma_v = 0.025$ S/m; relative dielectric values with $\varepsilon_r^h = 27$ and $\varepsilon_r^v = 7$; and relative magnetic permeability $\mu_r = 255$. The distance transmitter and receivers is 1.6 m as a typical induction T-R distance.

I do not use any kind of optimization techniques. However, the formulas require some gradient type measurement not only for magnetic, but also electric field.

In Figure 5, dotted and dashed lines show horizontal parameters on the first and in the middle panels. There are two formulas in Table 1 in the middle panel. When one use the upper formula, the result is negative. This is illustrated with dashed lines. The lower formula gives positive values. This is showed with dotted lines. As for the horizontal parameters, they are depicted with a solid line in Figure 5. As seen from Table 1, the real parts of the formulas give conductivity values. Dielectric values can be calculated from the imaginary parts of the formulas.

CONCLUSION

I have considered a hypothetical induction instrument in a uniaxial medium. In such a medium, one can calculate constitutive parameters such as conductivity, dielectric permittivity, and magnetic permeability. Nowadays, 3DEX and Rt Scanner technology is quit mature, but considered a suggested hypothetical instrument may be useful for formation evaluation since conductivity, dielectric permittivity and magnetic permeability values are important with respect to petrophysical parameters, because they are related to water saturation, porosity, and permeability. The formulas developed the present study are exact, since they have derived from Maxwell's equations. I have not used any kind of approximation in order to derive these parameters. Application of this method requires some gradient type field measurement, which might be the most difficult part of this application.

Appendix A

Equation (1) can explicitly be written

$$
\begin{bmatrix}
a_x & a_y & a_z \\
\dfrac{\partial}{\partial x} & \dfrac{\partial}{\partial y} & \dfrac{\partial}{\partial z} \\
H_x^x & H_y^x & H_z^x
\end{bmatrix}
$$

$$
=
\begin{bmatrix}
\sigma_h - i\omega\varepsilon_h & 0 & 0 \\
0 & \sigma_h - i\omega\varepsilon_h & 0 \\
0 & 0 & \sigma_v - i\omega\varepsilon_v
\end{bmatrix}
\begin{bmatrix}
E_x^x a_x \\
E_y^x a_y \\
E_z^x a_z
\end{bmatrix},
\tag{A.1}
$$

where a_x, a_y and a_z are the unit vectors in a Cartesian coordinate system. H and E are the components of corresponding fields either magnetic or electric. Superscripts stand for the magnetic moment direction, while subscripts stand for the field directions. The first, the electric and magnetic field components will be considered generated by an x-directed magnetic dipole. For this purpose, I rewrite Equation (A.1) and I have T

$$
\left(\frac{\partial H_z^x}{\partial y} - \frac{\partial H_y^x}{\partial z} \right) a_x = \left(\sigma_h - i\omega\varepsilon_h \right) E_x^x a_x
\tag{A.2}
$$

$$
\left(\frac{\partial H_z^x}{\partial x} - \frac{\partial H_x^x}{\partial z} \right) a_y = \left(\sigma_h - i\omega\varepsilon_h \right) E_y^x a_y
\tag{A.3}
$$

and

$$\left(\frac{\partial H_y^x}{\partial x} - \frac{\partial H_x^x}{\partial y}\right)a_z = \left(\sigma_v - i\omega\varepsilon_v\right)E_z^x a_z$$

(A.4)

Equations (A.2), (A.3) and (A.4) allow us to estimate horizontal and vertical conductivities from electric and magnetic fields. Continuing derivation yields

$$\sigma_h - i\omega\varepsilon_h = \frac{\left(\dfrac{\partial H_z^x}{\partial y} - \dfrac{\partial H_y^x}{\partial z}\right)}{E_x^x},$$

(A.5)

$$\sigma_h - i\omega\varepsilon_h = \frac{\left(\dfrac{\partial H_z^x}{\partial x} - \dfrac{\partial H_x^x}{\partial z}\right)}{E_y^x},$$

(A.6)

and

$$\sigma_v - i\omega\varepsilon_v = \frac{\left(\dfrac{\partial H_y^x}{\partial x} - \dfrac{\partial H_x^x}{\partial y}\right)}{E_z^x}.$$

(A.7)

The real part of Equations (A.5), (A.6) and (A.7) gives conductivity and the imaginary part of the same equation yields to permittivity. Continuing setting up Table 1, electric-magnetic field components generated by a y-directed magnetic dipole gives three more formulas. Let us look at Equation (1) again,

$$\begin{bmatrix} a_x & a_y & a_z \\ \dfrac{\partial}{\partial x} & \dfrac{\partial}{\partial y} & \dfrac{\partial}{\partial z} \\ H_x^y & H_y^y & H_z^y \end{bmatrix}$$

$$= \begin{bmatrix} \sigma_h - i\omega\varepsilon_h & 0 & 0 \\ 0 & \sigma_h - i\omega\varepsilon_h & 0 \\ 0 & 0 & \sigma_v - i\omega\varepsilon_v \end{bmatrix} \begin{bmatrix} E_x^y a_x \\ E_y^y a_y \\ E_z^y a_z \end{bmatrix},$$

(A.8)

and from Equation (A.8) I can write as the following equations,

$$\left(\frac{\partial H_z^y}{\partial y} - \frac{\partial H_y^y}{\partial z}\right)a_x = \left(\sigma_h - i\omega\varepsilon_h\right)E_x^y a_x,$$

(A.9)

$$\left(\frac{\partial H_z^y}{\partial x} - \frac{\partial H_x^y}{\partial z} \right) a_y = \left(\sigma_h - i\omega\varepsilon_h \right) E_y^y a_y$$

(A.10)

and

$$\left(\frac{\partial H_y^y}{\partial x} - \frac{\partial H_x^y}{\partial y} \right) a_z = \left(\sigma_h - i\omega\varepsilon_h \right) E_z^y a_z$$

(A.11)

Then parameters can be estimated as the following expressions

$$\sigma_h - i\omega\varepsilon_h = \frac{\left(\frac{\partial H_z^y}{\partial y} - \frac{\partial H_y^y}{\partial z} \right)}{E_x^y}$$

(A.12)

$$\sigma_h - i\omega\varepsilon_h = \frac{\left(\frac{\partial H_z^y}{\partial x} - \frac{\partial H_x^y}{\partial z} \right)}{E_y^y}$$

(A.13)

and

$$\sigma_v - i\omega\varepsilon_v = \frac{\left(\frac{\partial H_y^y}{\partial x} - \frac{\partial H_x^y}{\partial y} \right)}{E_z^y}$$

(A.14)

Conductivities can be calculated from the real part of Equations (A.12), (A.13) and (A.14) and the imaginary part of the same equation may be used for estimating dielectric constant.

Further derivation using a z-directed magnetic dipole gives three more formulas. Let continue the derivation from Equation (1),

$$\begin{vmatrix} a_x & a_y & a_z \\ \dfrac{\partial}{\partial x} & \dfrac{\partial}{\partial y} & \dfrac{\partial}{\partial z} \\ H_x^z & H_y^z & H_z^z \end{vmatrix}$$

$$= \begin{bmatrix} \sigma_h - i\omega\varepsilon_h & 0 & 0 \\ 0 & \sigma_h - i\omega\varepsilon_h & 0 \\ 0 & 0 & \sigma_v - i\omega\varepsilon_v \end{bmatrix} \begin{bmatrix} E_x^z a_x \\ E_y^z a_y \\ E_z^z a_z \end{bmatrix}$$

(A.15)

Proceeding with some very simple algebra allows us to have other parameters. From Equation (A.15) it is very easy to have

$$\left(\frac{\partial H_z^x}{\partial y} - \frac{\partial H_y^x}{\partial z} \right) a_x = \left(\sigma_h - i\omega\varepsilon_h \right) E_x^x a_x$$

(A.16)

$$\left(\frac{\partial H_z^x}{\partial x} - \frac{\partial H_x^x}{\partial z} \right) a_y = \left(\sigma_h - i\omega\varepsilon_h \right) E_y^x a_y$$

(A.17)

and

$$\left(\frac{\partial H_y^x}{\partial x} - \frac{\partial H_x^x}{\partial y} \right) a_z = \left(\sigma_v - i\omega\varepsilon_v \right) E_z^x a_z$$

(A.18)

Again, having conductivities and dielectric permittivites are easy. From the last three equations I can derive as the following expressions:

$$\sigma_h - i\omega\varepsilon_h = \frac{\left(\dfrac{\partial H_z^z}{\partial y} - \dfrac{\partial H_y^z}{\partial z} \right)}{E_x^z}$$

(A.19)

$$\sigma_h - i\omega\varepsilon_h = \frac{\left(\dfrac{\partial H_z^z}{\partial x} - \dfrac{\partial H_x^z}{\partial z} \right)}{E_y^z}$$

(A.20)

and

$$\sigma_v - i\omega\varepsilon_v = \frac{\left(\dfrac{\partial H_y^z}{\partial x} - \dfrac{\partial H_x^z}{\partial y} \right)}{E_z^z}$$

(A.21)

From Equations (A.5), (A.6), (A.7), (A.8), (A.9), (A.10), (A.12), (A.13), and (A.14) apparent conductivity and permittivity can be calculated. For this purpose, the real part of the Equations (A.19) and (A.20) are used. Horizontal and vertical conductivities may be estimated with

$$\sigma_h = \Re \left(\frac{\left(\dfrac{\partial H_z^z}{\partial y} - \dfrac{\partial H_y^z}{\partial z} \right)}{E_x^z} \right)$$

(A.22)

$$\sigma_h = \Re \left(\frac{\left(\dfrac{\partial H_y^z}{\partial x} - \dfrac{\partial H_x^z}{\partial y} \right)}{E_z^z} \right).$$

(A.23)

Similar to previous two formulas, apparent dielectric permittivities can be

$$\varepsilon_h = \Im \left(\frac{\left(\dfrac{\partial H_z^z}{\partial y} - \dfrac{\partial H_y^z}{\partial z} \right)}{i\omega E_x^z} \right),$$

(A.24)

and

$$\varepsilon_v = \Im \left(\frac{\left(\dfrac{\partial H_y^z}{\partial x} - \dfrac{\partial H_x^z}{\partial y} \right)}{i\omega E_z^z} \right).$$

(A.25)

Further, consider relative permittivity as $\varepsilon_h = \varepsilon_h^r \varepsilon_0$ and $\varepsilon_v = \varepsilon_v^r \varepsilon_0$ and rewrite the last two equations, I have

$$\varepsilon_h^r = \Im \left(\frac{\left(\dfrac{\partial H_z^z}{\partial y} - \dfrac{\partial H_y^z}{\partial z} \right)}{i\varepsilon_0 \omega E_x^z} \right)$$

(A.26)

and

$$\varepsilon_v^r = \Im \left(\frac{\left(\dfrac{\partial H_y^z}{\partial x} - \dfrac{\partial H_x^z}{\partial y} \right)}{i\varepsilon_0 \omega E_z^z} \right).$$

(A.27)

Appendix B

The formulas for the magnetic permeability estimation can be derived by using Equation (2). Rewrite Equation (2) explicitly with x directed a magnetic dipole. Bear in mind that the magnetic permeability is scalar. I can write it as

$$
\begin{bmatrix}
a_x & a_y & a_z \\
\dfrac{\partial}{\partial x} & \dfrac{\partial}{\partial y} & \dfrac{\partial}{\partial z} \\
E_x^x & E_y^x & E_z^x
\end{bmatrix}
= i\omega\mu
\begin{bmatrix}
H_x^x a_x \\
H_y^x a_y \\
H_z^x a_z
\end{bmatrix}
+ i\omega\mu
\begin{bmatrix}
M^x a_x \\
M^x a_y \\
M^x a_z
\end{bmatrix}
$$

(A.28)

The first component of Equation (A.28) with x-directed magnetic dipole with M a unit moment, from the previous step I can proceed with

$$
\left(\frac{\partial E_z^x}{\partial y} - \frac{\partial E_y^x}{\partial z} \right) a_x = i\omega\mu H_x^x a_x + i\omega\mu M^x a_x
$$

(A.29)

and keep on derivation, which yields

$$
i\omega\mu = \frac{\left(\dfrac{\partial E_z^x}{\partial y} - \dfrac{\partial E_y^x}{\partial z} \right)}{M^x + H_x^x}
$$

(A.30)

From Equation (A.30), it is easy to have magnetic permeability:

$$
\mu = \frac{1}{i\omega} \frac{\left(\dfrac{\partial E_z^x}{\partial y} - \dfrac{\partial E_y^x}{\partial z} \right)}{M^x + H_x^x}
$$

(A.31)

It can be derived some formulas for the magnetic permeability by using the same procedure as on the previous component derivation.

$$
\left(\frac{\partial E_z^x}{\partial x} - \frac{\partial E_x^x}{\partial z} \right) a_y = i\omega\mu H_y^x a_y + i\omega\mu M^x a_y
$$

(A.32)

$$
i\omega\mu = \frac{\left(\dfrac{\partial E_z^x}{\partial x} - \dfrac{\partial E_x^x}{\partial z} \right)}{M^x + H_y^x}
$$

(A.33)

$$\mu = \frac{1}{i\omega} \frac{\left(\dfrac{\partial E_z^x}{\partial x} - \dfrac{\partial E_x^x}{\partial z}\right)}{M^x + H_y^x} .$$

(A.34)

From the last component of the corresponding vector, one can get

$$\left(\frac{\partial E_y^x}{\partial x} - \frac{\partial E_x^x}{\partial y}\right) a_z = i\omega\mu H_z^x a_z + i\omega\mu M^x a_z .$$

(A.35)

$$i\omega\mu = \frac{\left(\dfrac{\partial E_y^x}{\partial x} - \dfrac{\partial E_x^x}{\partial y}\right)}{M^x + H_z^x} ,$$

(A.36)

and

$$\mu = \frac{1}{i\omega} \frac{\left(\dfrac{\partial E_y^x}{\partial x} - \dfrac{\partial E_x^x}{\partial y}\right)}{M^x + H_z^x} .$$

(A.37)

Then, carry on the derivation

$$\begin{vmatrix} a_x & a_y & a_z \\ \dfrac{\partial}{\partial x} & \dfrac{\partial}{\partial y} & \dfrac{\partial}{\partial z} \\ E_x^y & E_y^y & E_z^y \end{vmatrix} = i\omega\mu \begin{bmatrix} H_x^y a_x \\ H_y^y a_y \\ H_z^y a_z \end{bmatrix} + i\omega\mu \begin{bmatrix} M^y a_x \\ M^y a_y \\ M^y a_z \end{bmatrix} ,$$

(A.38)

$$\left(\frac{\partial E_z^y}{\partial y} - \frac{\partial E_y^y}{\partial z}\right) a_x = i\omega\mu H_x^y a_x + i\omega\mu M^y a_x ,$$

(A.39)

$$i\omega\mu = \frac{\left(\dfrac{\partial E_z^y}{\partial y} - \dfrac{\partial E_y^y}{\partial z}\right)}{M^y + H_x^y} ,$$

(A.40)

yields one more formula for the magnetic permeability of the medium, which is

$$\mu = \frac{1}{i\omega} \frac{\left(\dfrac{\partial E_z^y}{\partial y} - \dfrac{\partial E_y^y}{\partial z} \right)}{M^y + H_x^y}.$$

(A.41)

Considering the y component of the vector, then keep on

$$\left(\frac{\partial E_z^y}{\partial x} - \frac{\partial E_x^y}{\partial z} \right) a_y = i\omega\mu H_y^y a_y + i\omega\mu M^y a_y$$

(A.42)

which yields

$$i\omega\mu = \frac{\left(\dfrac{\partial E_z^y}{\partial x} - \dfrac{\partial E_x^y}{\partial z} \right)}{M^y + H_y^y},$$

(A.43)

then it is easy to have

$$\mu = \frac{1}{i\omega} \frac{\left(\dfrac{\partial E_z^y}{\partial x} - \dfrac{\partial E_x^y}{\partial z} \right)}{M^y + H_y^y}.$$

(A.44)

When I apply similar procedure, the rest of the magnetic formulas may be derived:

$$\left(\frac{\partial E_y^y}{\partial x} - \frac{\partial E_x^y}{\partial y} \right) a_z = i\omega\mu H_z^y a_z + i\omega\mu M^y a_z,$$

(A.45)

$$i\omega\mu = \frac{\left(\dfrac{\partial E_y^y}{\partial x} - \dfrac{\partial E_x^y}{\partial y} \right)}{M^y + H_z^y},$$

(A.46)

$$\mu = \frac{1}{i\omega} \frac{\left(\dfrac{\partial E_y^y}{\partial x} - \dfrac{\partial E_x^y}{\partial y} \right)}{M^y + H_z^y}.$$

(A.47)

$$\begin{vmatrix} a_x & a_y & a_z \\ \dfrac{\partial}{\partial x} & \dfrac{\partial}{\partial y} & \dfrac{\partial}{\partial z} \\ E_x^z & E_y^z & E_z^z \end{vmatrix} = i\omega\mu \begin{bmatrix} H_x^z a_x \\ H_y^z a_y \\ H_z^z a_z \end{bmatrix} + i\omega\mu \begin{bmatrix} M^z a_x \\ M^z a_y \\ M^z a_z \end{bmatrix},$$

(A.48)

$$\left(\frac{\partial E_z^z}{\partial y} - \frac{\partial E_y^z}{\partial z}\right)a_x = i\omega\mu H_x^z a_x + i\omega\mu M^z a_x \tag{A.49}$$

$$i\omega\mu = \frac{\left(\dfrac{\partial E_z^z}{\partial y} - \dfrac{\partial E_y^z}{\partial z}\right)}{M^z + H_x^z} \tag{A.50}$$

$$\mu = \frac{1}{i\omega}\frac{\left(\dfrac{\partial E_z^z}{\partial y} - \dfrac{\partial E_y^z}{\partial z}\right)}{M^z + H_x^z}. \tag{A.51}$$

$$\left(\frac{\partial E_z^z}{\partial x} - \frac{\partial E_x^z}{\partial z}\right)a_y = i\omega\mu H_y^z a_y + i\omega\mu M^z a_y \tag{A.52}$$

$$i\omega\mu = \frac{\left(\dfrac{\partial E_z^z}{\partial x} - \dfrac{\partial E_x^z}{\partial z}\right)}{M^z + H_y^z} \tag{A.53}$$

$$\mu = \frac{1}{i\omega}\frac{\left(\dfrac{\partial E_z^z}{\partial x} - \dfrac{\partial E_x^z}{\partial z}\right)}{M^z + H_y^z}. \tag{A.54}$$

$$\left(\frac{\partial E_y^z}{\partial x} - \frac{\partial E_x^z}{\partial y}\right)a_z = i\omega\mu H_z^z a_z + i\omega\mu M^z a_z \tag{A.55}$$

$$i\omega\mu = \frac{\left(\dfrac{\partial E_y^z}{\partial x} - \dfrac{\partial E_x^z}{\partial y}\right)}{M^z + H_z^z}, \tag{A.56}$$

and

$$\mu = \frac{1}{i\omega}\frac{\left(\dfrac{\partial E_y^z}{\partial x} - \dfrac{\partial E_x^z}{\partial y}\right)}{M^z + H_z^z}. \tag{A.57}$$

218   Handbook of Electrical Measurements

REFERENCES

1. C. A. Balanis, "Advanced Engineering Electromagnetics," John Wiley & Sons, Inc., Hoboken, 1989.

2. R. S. Carmichael, "Practical Handbook of Physical Properties of Rocks and Minerals," CRC, Boca Raton, 1989.

3. F. S. Grant and G. F. West, "Interpretation Theory in Applied Geophysics," McGraw-Hill Book Company, New York, 1965.

4. M. N. Nabighian, Ed., "Electromagnetic Methods in Applied Geophysics V. I: Theory," SAGE Publication, Thousand Oaks, 1987.

5. M. S. Zhdanov and G. V. Keller, "The Geoelectrical Methods in Geophysical Exploration," Elsevier, Amsterdam, 1994.

6. J. D. Klein, P. R. Martin and D. F. Allen, "The Petrophysics of Electrically Anisotropic Reservoirs," SPWLA 36th Annual Logging Symposium, Paris, 26-29 June 1995, Paper HH.

7. M. S. Zhdanov, D. Kennedy and E. Peksen, "Foundations of Tensor Induction Well-Logging," Petrophysics, Vol. 42, No. 6, 2001, pp. 588-610.

8. Z. Zhang, L. Yu, B. Kriegshauser and L. Tabarovsky, "Determination of Relative Angles and Anisotropic Resistivity Using Multicomponent Induction Logging Data," Geophysics, Vol. 69, No. 4, 2004, pp. 898-908. doi:10.1190/1.1778233

9. B. I. Anderson, T. D. Barber and M. G. Lulling, "The Response of Induction Tools to Dipping Anisotropic Formations," SPWLA 36th Annual Logging Symposium, Paris, 26-29 June 1995, Paper D.

10. M. G. Lüling, R. Rosthal and F. Shray, "Processing and Modeling 2 MHz Tools in Dipping, Laminated Anisotropic Formations," SPWLA 35th Annual Logging Symposium, Tulsa, 19-22 June 1994, Paper QQ.

11. J. H. Moran and S. C. Gianzero, "Effects of Formation Anisotropy on Resistivity Logging Measurements," Geophysics, Vol. 44, No. 7, 1979, pp. 1266-1286.doi:10.1190/1.1441006

12. L. Zhong, C. L. Shen, R. Liu, M. Bittar and G. Hu, "Simulation of Tri-Axial Induction Logging Tools in Layered Anisotropic Dipping Formations," 76th Annual International Meeting SEG, New Orleans, 2006, pp. 456- 460.

13. L. L. Zhong, J. Li, L. C. Shen and R. C. Liu, "Computation of Triaxial Induction Logging Tools in Layered Anisotropic Dipping Formations," IEEE Transactions on Geoscience and Remote Sensing, Vol. 46, No. 4, 2008, pp. 1148-1163. doi:10.1109/TGRS.2008.915749.

14. S. Gianzero, D. Kennedy, L. Gao and L. San Martin, "The Response of a Triaxial Sonde in a Biaxial Anisotropic Medium," Petrophysics, Vol. 43, No. 3, 2002, pp. 172- 184.

15. A. G. Nekut, "Anisotropy Induction Logging," Geophysics, Vol. 59, No. 3, 1994, pp. 345-350. doi:10.1190/1.1443596

16. N. Yuan, X. C. Nie, R. Liu and C. W. Qiu, "Simulation of Full Responses of a Triaxial Induction Tool in a Homogeneous Biaxial Anisotropic Formation," Geophysics, Vol. 75, No. 2, 2010, pp. E101-E114. doi:10.1190/1.3336959

17. A. Gribenko and M. S. Zhdanov, "Rigorous 3D Inversion of Tensor Electrical and Magnetic Induction Well Logging Data Inhomogeneous Media," 79th Annual International Meeting SEG, Houston, 2009, pp. 431-435.

18. M. S. Zhdanov, "Method and Apparatus for Gradient Electromagnetic Induction Well Logging," PCT WO 2005/ 083467 A1. 2005.

19. X. Lu, D. L. Alumbaugh and C. J. Weiss, "The Electric Fields and Currents Produced by Induction Logging Instruments in Anisotropic Media," Geophysics, Vol. 67, No. 2, 2002, pp. 478-483. doi:10.1190/1.1468607

20. E. Peksen, "Methods for Interpretation of Tensor Induction Well Logging in Layered Anisotropic Formations," Ph.D. Thesis, University of Utah, Salt Lake City, 2004.

21. T. Wang, "The Electromagnetic Smoke Ring in a Transversely Isotropic Medium," Geophysics, Vol. 67, No. 6, 2002, pp. 1779-1789.

22. M. Rabinovich, L. Tabarovsky, B. Corley, L. van der Horst and M. Epov, "Processing Multi-Component Induction Data for Formation Dips and Anisotropy," Petrophysics, Vol. 47, No. 6, 2006, pp. 506-526. doi:10.1190/1.1527078

Chapter 10

RETINAL VESSEL WIDTH MEASUREMENT AT BRANCHINGS USING AN IMPROVED ELECTRIC FIELD THEORY-BASED GRAPH APPROACH

Xiayu Xu[1] , Joseph M. Reinhardt[1] , Qiao Hu[2] , Benjamin Bakall[3] , Paul S. Tlucek[3] , Geir Bertelsen[5] , Michael D. Abra`moff[1,2,3,4]

[1]Department of Biomedical Engineering, University of Iowa, Iowa City, Iowa, United States of America

[2]Department of Electrical and Computer Engineering, University of Iowa, Iowa City, Iowa, United States of America

[3]Department of Ophthalmology and Visual Science, University of Iowa, Iowa City, Iowa, United States of America

[4]Veteran's Administration Medical Center, Iowa City, Iowa, United States of America

[5]Department of Community Medicine, University of Tromsø, Tromsø, Norway

ABSTRACT

The retinal vessel width relationship at vessel branch points in fundus images is an important biomarker of retinal and systemic disease. We propose a fully automatic method to measure the vessel widths at branch points in fundus images. The method is a graph-based method, in which a graph construction method based on electric field theory is applied which specifically deals with complex branching patterns. The vessel centerline image is used as the initial segmentation of the graph. Branching points are detected on the vessel centerline image using a set of detection kernels. Crossing points are distinguished from branch points and excluded. The electric field based graph method is applied to construct the graph. This method is inspired by the non-intersecting force lines in an electric field. At last, the method is further improved to give a consistent vessel width measurement for the whole vessel tree. The algorithm was validated on 100 artery branchings and 100 vein branchings selected from 50 fundus images by comparing with vessel width measurements from two human experts.

INTRODUCTION

Motivation

The retinal vessel width relationship at vessel branch points in fundus images is an important risk factor of retinal and systemic disease, including ischemic heart disease, hypertension and brain abnormalities, [1], [2]. The relationship between retinal arterial diameters at branch points conform to predicted optimal values in normal subjects, but deviate significantly in patients with peripheral vascular disease [3]. Studies also showed that this relationship deviates from the theoretic optimum with advancing age [4]. Increased branching coefficients of retinal vessels have been reported to be associated with periventricular white matter hyperintensities and ischaemic heart disease, and decreased branching coefficient with deep white matter hyperintensities [1]. In some conditions, such as hypertension, the smaller arteries are affected more than the larger ones [2]. Compared with using vessel width directly as a parameter, the vessel width relationship at the branch point is dimensionless, which allows measurements without correcting for the differences in magnification by the optics of the eye across individuals, caused by different refractive errors. [5], [6].

Previous Work

Though image analysis of retinal vessel has been widely studied and over two hundreds papers have been published in the field of retinal vasculature detection and vessel width measurements [7], only a few studies focus on vessel width measurement at branch point and proposed specific methods to solve these type of problems [1], [3], [8]–[11]. In general, studies that treat branch points separately focused on two phases: a) detection of branch point, and b) vessel width measurement at branch point.

Most of the retinal branch point detection methods target using the bifurcations as a landmark for further image analysis [8]–[10], [12]–[14], such as retinal image registration. Shen et al. in 2001, proposed a real-time landmark extraction method from fundus images. This model-based method detects branch points as a fragment of the vasculature that consist of two relatively straight anti-parallel edges with either an intensity peak or an intensity valley in between. Tsai et al. further refined this method in 2004, in which the detected branch point is used as the initial estimated branch point. Then an exclusion region is provided around the estimated branch point and the location is further refined within the exclusion region. In 2008, Bhuiyan *et al.* proposed a method to detect vascular bifurcations and crossovers based on the vessel geometrical features. A binary vessel image is first segmented from the color fundus image and morphological thinning operation is applied to find the vessel centerline. Subsequently, rotational invariant 3×3 masks are used to detect potential bifurcations and crossover points. Finally, the geometrical and topological properties are used to refine the result. A detection accuracy of 95.82% was reported.

Compared with vessel branch point detection, fewer studies have addressed the problem of width measurement at branch points. In most retinal branching studies, the vessel width is measured by manual methods ([2], [4], [15], [16]) or semi-automatic methods ([1], [3], [11]). In a study on the relationship of peripheral vascular disease and arterial bifurcation diameter, Chapman *et al.* proposed a semi-automatic method to measure the vessel width for arteries[3]. This method needs human expert operators to draw lines perpendicular to arteries and then an automatic method is used to determine the points of maximum intensity variation based on the cross-sectional profiles [17]. This method is not specific for retinal vascular branchings. Doubal *et al.* proposed a semi-automatic method that can track down each branch if the branch center is identified by a trained grader. Then the cross-sectional profiles can be obtained and a Gaussian curve is fit to determine the width [1]. Each profile needs manual inspection after the fitting. Patton *et al.* used a similar semi-automatic method to calculate the branching vessel width [11]. The human operator manually identifies the arterial and venous branch points and draw a line perpendicular to the vessel. Four other intensity profiles can be automatically generated from the given profile. Each profile can be rejected or accepted by the operator. Then a Gaussian fitting is used to identify the vessel width.

To our knowledge, no fully automatic method has been developed to specifically deal with the vessel width measurement problem at branch points in fundus image.

METHODOLOGY

We have previously published a graph-based method to measure the vessel width for straight vessel segments [18]. The vessel centerline image is used to obtain the base nodes for the graph. Then, a two-slice three-dimensional graph is luxated for each vessel segment. The graph columns are built perpendicular to the vessel growing directions. The perpendicular direction is calculated using principal component analysis (PCA). A smoothness constraint between the two slices is applied. Thus, the simultaneous two-dimensional boundary segmentation problem is transformed into a three-dimensional surface segmentation problem. It is further converted to a minimum closed set problem in a node-weighted graph. By solving this minimum closed set problem, the two boundaries of blood vessels can be determined and the vessel width can be measured. Because graph columns are constructed along the second principal component of PCA, this method will be referred to as the PCA-method in the remainder of this paper.

The PCA-method for straight vessel width measurement cannot be extended to the measurement of branch point directly. First, the graph columns will intersect each other as the graph columns approach the branching center, resulting in multiple vessel width measurements at the same point. Second, the graph columns might run into adjoining vessel branches, resulting in meaningless measurements of vessel width inside the vessel segment. In order to address these problems, a different graph construction method is needed. We propose an electric field theory motivated graph construction method to solve this problem. Similarly, because the graph columns are constructed along the electric lines of force, this method is referred to as the ELF-method.

Pre-processing and Bifurcation Detection

The goal of pre-processing is to obtain a vessel centerline image as the initial segmentation. We start with the vessel segmentation map as proposed by Niemeijer *et al.* in [19]. As in the original study of the vessel segmentation, we used the images and reference standard of the DRIVE database (http://www.isi. uu.nl/ Research/Databases/DRIVE/). One example image is shown in Figure 1 (b). The vessel segmentation map is a gray scale image with each pixel assigned the likelihood of being in a blood vessel, the higher the intensity, the higher the likelihood. By thresholding the vessel segmentation map, a binary vessel segmentation image is generated. A constant low threshold of 70 is chosen to better maintain the continuity of blood vessels. The trade-off is that small regions of noise may not be suppressed adequately. Thus, vessel regions with an area smaller than 20 pixels are erased from the binary vessel image. A sequential thinning approach is then applied to the binary vessel segmentation image to find the vessel centerline [20].

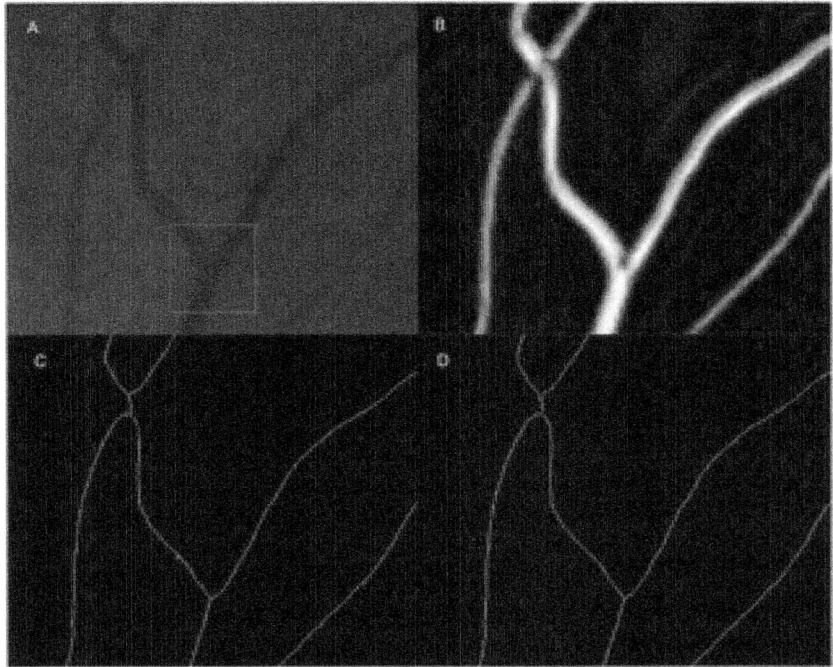

Figure 1: Spur pruning and crossing point exclusion on vessel centerline image.

(a) An enlarged part of a color fundus image. Branch inside the rectangle indicates the branch in Figure 1. (b) The corresponding vesselness map. (c) The corresponding vessel centerline image with spur presented. A crossing point is split into two branch points. (d) The corresponding vessel centerline image after spur pruning and crossing point exclusion.

doi:10.1371/journal.pone.0049668.g001

Branch points are detected on the vessel centerline image using a series of 3×3 kernels [20]. The kernels are given in Equation 1.

$$k_1 = \begin{bmatrix} 1 & 0 & 1 \\ 0 & 1 & 0 \\ 0 & 0 & 1 \end{bmatrix}, k_2 = \begin{bmatrix} 1 & 0 & 1 \\ 0 & 1 & 0 \\ 0 & 1 & 0 \end{bmatrix}, \cdots, k_{16} = \begin{bmatrix} 0 & 0 & 1 \\ 0 & 1 & 0 \\ 1 & 0 & 1 \end{bmatrix} \tag{1}$$

These kernels can effectively detect all branch points which are shown on the vessel centerline image. The ability for the proposed algorithm to automatically localize the correct vessel centerline was validated in Reference [18]. However, not all detected branch points are true vessel branchings, as shown in Figure 1 (c). False branches can result from abrupt vessel width changes, vessel direction changes, or from noise. Moreover, the vessel crossings are often detected as two adjacent branchings on the vessel centerline image. Hence, a spur pruning and crossing point exclusion step is applied.

Starting from a detected branch point, all three branches are traced. If another branch point is reached within certain distance, the two branch points are regarded as a single crossing point and excluded from further study. On the other hand, if an end point is reached within certain distance, the traced branch is regarded as a spur and removed from the centerline image.

Graph-Based Vessel Boundary Segmentation

We apply an electric field theory motivated graph construction method to build the graph at branch points. This graph construction method was first proposed by Yin *et al.* in 2009 [21]. The method is inspired by the non-intersecting property of electric lines of force. Recall Coulomb›s law:

$$E_i = \frac{1}{4\pi\varepsilon_0} \frac{Q}{r^2} \hat{r},$$

(2)

where E_i is the electric field at point i, Q is the charge of point i, r is the distance from the point i to the evaluation point j, $\hat{r}$ is the unit vector pointing from the point i to the evaluation point j, and ε_0 is the vacuum permittivity. The total electric field E at point j is the sum of E_i:

$$E = \sum E_i,$$

(3)

the electric field has the same direction as the *electric line of force* (ELF).

When multiple source points exist in an electric field, the electric line of force holds a non-intersecting property. If we change r^2 to $r^n (n > 0)$, the non-intersection property still holds. The difference is that the vertices with larger distances will be penalized in ELF computing. A value of $n = 4$ is used to decrease the effect from pixels with a larger distance and hence increase the robustness of local ELF computation.

By applying this theory to the problem of graph construction at branch points, we assume each vessel centerline pixel is a positive unit charge. The electric line of force is calculated and the graph is constructed along the electric line of force, as illustrated in Figure 2 (b).

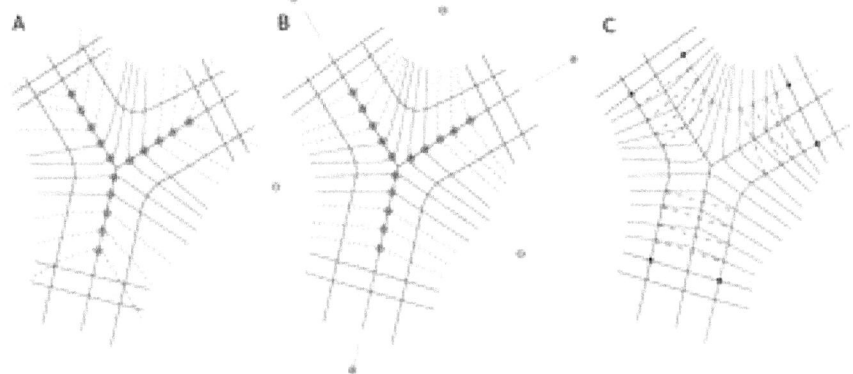

Figure 2: Illustration of problems in applying traditional graph construction method to branch points and introduction to electric field theory based graph construction method.

(a) A figure to illustrate possible problems in applying traditional graph construction method to branch point situation. The problems include possible graph column intersection and graph columns running into another vessel branch. (b) Introduction to electric field theory based graph construction method. Its most attractive is the non-intersecting property of electric field lines.

doi:10.1371/journal.pone.0049668.g002

If we consider branch points to be isolated from vessel trees, the graph given above is good enough to give a reliable measurement. However, if a consistent vessel width measurement is desired for the whole vessel tree, problems will arise at the transition from the graph built using ELF-method for branch points to the graph built using PCA-method for the adjoining straight vessels, as shown in Figure 3 (a). The electric line of force points outside at the end of each branch. When the two different types of graphs are connected, the graph columns from different methods will intersect.

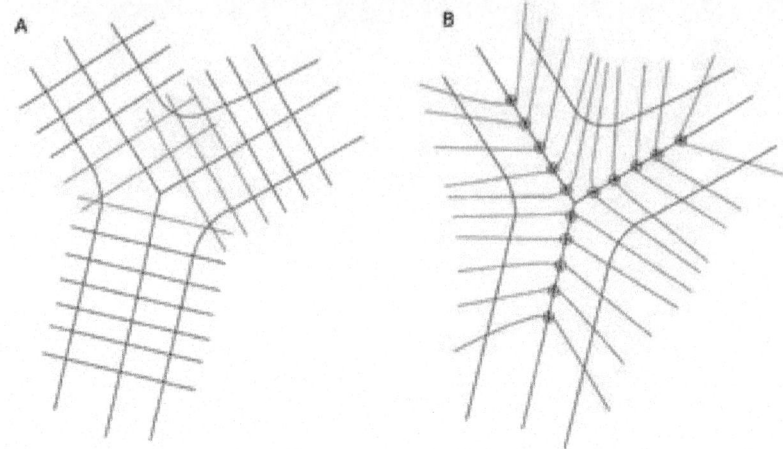

Figure 3: Improvement of the electric field theory based graph construction method.

For illustration purposes, not all graph columns are shown in the figure. (a) Problems encountered when combining the traditional graph and electric field theory based graph. The graph columns intersect each other because the graph columns from the electric field theory based method point outside rather than the normal direction when approaching the end of branches, as shown in dashed red lines. (b) Improved combined graph. Two strategies are introduced to improve the construction of the combined graph. (c) Three graphs are built for the three boundaries separately, as shown in color green, blue, and purple. After the three boundaries are determined, the vessel width is measured as the Euclidean distance between the corresponding nodes from the same centerline pixel.

doi:10.1371/journal.pone.0049668.g003

We propose two strategies to improve the electric field theory based graph construction method, as illustrated in Figure 3 (b). First, in order to pull the electric lines of force towards the normal direction at the end of each branch, extra positive charges and negative charges are added in the electric field. Positive charges with extremely large values are positioned at the infinite extension of each branch. This is used to generate a parallel force along the vessel growing direction that can push the graph columns toward the perpendicular direction. A distance of 10,000 pixels is used to simulate the infinite distance and the charge value was chosen to be 100,000 unit charges. Values of the distance and charge are not sensitive and do not have a large effect as long as they are big enough to be considered as "infinity" in this case. In order to further pull the graph columns towards the normal direction, three negative charges are positioned in the middle points of three branches. Values of the negative charges are also not sensitive as long as they are comparable to the value of adjacent positive charges. We chose the value to be five in this case.

Second, a combined graph construction method is introduced at the transition location. Six centerline pixels at the transition location are used to build the combined graph. First of all, the location of graph nodes are calculated using both the PCA-method and the ELF-method, given by $(x_{pca|i}, y_{pca|i})$ and $(x_{elf|i}, y_{elf|i})$, where $i = 0,1,...,5$ is the distance from the graph column to the branching center. Then the node locations in the combined graph are calculated using a weighted linear interpolation of $(x_{pca|i}, y_{pca|i})$ and $(x_{elf|i}, y_{elf|i})$:

$$\begin{cases} x_i = 0.2ix_{pca|i} + (1 - 0.2i)x_{elf|i} \\ y_i = 0.2iy_{pca|i} + (1 - 0.2i)y_{elf|i} \end{cases}, \quad (4)$$

where $i = 0,1,...,5$.

Cost Function

The cost function is generated from convolution of the image's green channel with an oriented first order derivative of Gaussian. The green channel has been reported to show the highest contrast between the blood vessels and background [22]. A steerable first order derivative of Gaussian filter is used to implement the kernel [23]. The steerable filter has high responses to the gradient along different angles at different locations. The separable first order derivative of Gaussian along the x-axis and along the y-axis are given by G_1^0 and $G_1^{90°}$. The first order derivative of Gaussian along any angle θ is defined in Equation 5.

$$G_1^\theta = \cos(\theta)G_1^{0°} + \sin(\theta)G_1^{90°} \quad (5)$$

The original image is first convolved with G_1^0 and $G_1^{90°}$ to get the first order derivative image along the x-axis and y-axis. Then within each normal profile, the weights of the profile nodes are calculated according to equation 5. In our implementation, angle θ is the graph column direction, i.e., the direction of the electric lines of force at each graph node location.

Graph Search and Boundary Determination

For each branch point, the three boundaries are constructed as three independent single slice graphs. Intra-column smoothness constraints are set to maintain the smoothness within the slice. After the graphs are built, the optimal segmentation is found and the vessel boundary is determined as described by Li et al. [24].

Once the three boundaries are determined, the vessel width is measured as the Euclidean distance between the graph nodes on the segmentation from the same centerline pixel, as illustrated in Figure 3 (c).

Experimental Methods

A set of 100 artery branchings and 100 vein branchings were selected from 50 fundus images from 50 normal subjects and used to assess vessel width measurement performance (available on INSPIRE website: http://webeye. ophth.uiowa.edu/component/k2/item/270). These 50 normal subjects were selected randomly from the Tromsø study cohort. The Tromsø study was initiated in 1974 in an attempt to help combat the high mortality of cardiovascular diseases in Norway. The Tromsø study consists of six surveys that have been conducted in the municipality of Tromsø from 1974 to 2008. This population study includes 40051 subjects in total who have attended at least on of the six surveys. There are currently some 100 different ongoing research projects based on the data from the consecutive six surveys. A good overview of the Tromsø study is in Jacobsen et al [25]. All subjects provided written informed consent for participation in this study, and all images were de-identified before sharing with the University of Iowa. The research team at the University of Iowa did not have access to any patient identifiable information, and the study was therefore declared exempt by the institution review board of the University of Iowa. All research was in accordance with the tenets of the Declaration of Helsinki. The image resolution is 2196×1958 pixels, and images were stored in DICOM format. A pixel is approximately 3.7 μm on each side. For each fundus image, two artery branches and two vein branches were initially selected (based on their contrast) to evaluate the algorithm. In order to make sure the measurements given by experts and algorithm are comparable, i.e., at approximately the same location, for each branch, a start point and an end point were given to indicate where the measurements should be performed. For each branch, the start point was at approximately one vessel diameter (15 pixels) away from the branch center. The end point was at approximately two vessel diameters away from the branch center. The three start points and three end points are shown in Figure 4 as blue dots on the branch.

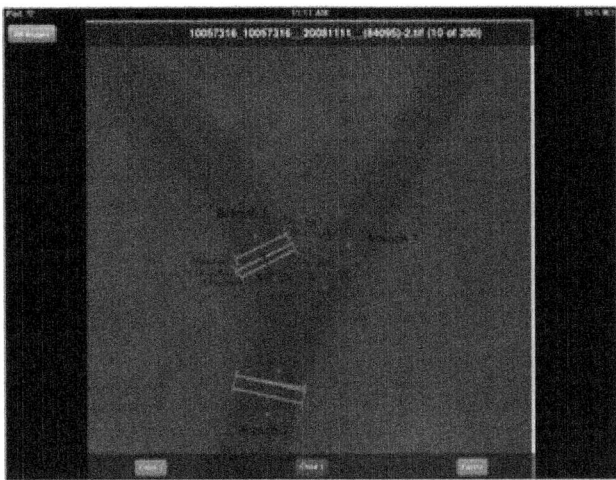

Figure 4: An illustration of the human expert annotation.

Blue dots denote the region where the measurement should be given. Black text were superimposed by the author for the sake of illustration. Vessel width for branch 1 is calculated as the average of the three width profiles. The branch center for branch 1 is calculated as the average of the three width profile center.

doi:10.1371/journal.pone.0049668.g004

The two experts are denoted as E_1 and E_2 respectively. The manual measurements were performed on a tablet-based color fundus image evaluation system ([26]) and the interface is given in Figure 4. The blue dots denote the region within which the measurement should be given. The black texts were superimposed by the author for the sake of illustration. Each expert was told to annotate at least three measurement profiles to each of the three branches for each selected branch point. Each measurement profile contains seven outputs: image name, image index, branch index, and profile start and end locations x_1, y_1, x_2, and y_2. The branch center is defined as $(x_k, y_k) = (\frac{1}{2n} \sum_{i=1}^{n} (x_{1i} + x_{2i}), \frac{1}{2n} \sum_{i=1}^{n} (y_{1i} + y_{2i}))$ and the width of the branch is defined as $w_k = \frac{1}{n} \sum_{i=1}^{n} \sqrt{(x_{1i} - x_{2i})^2 + (y_{1i} - y_{2i})^2}$, where $i = 1,...,n; (n \geq 3)$ is the number of measurement at the given branch and $k = 1,2$ denotes the two experts.

RESULTS

Typical vessel width measurements are shown in Figure 5. Red lines denote the vessel width measurement for branch points. Black lines denote the vessel width measurement for straight vessels.

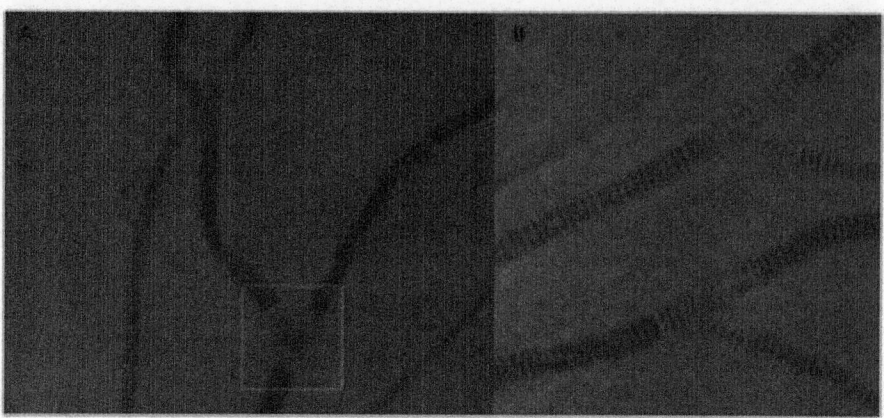

Figure 5: Typical vessel width measurement images, with red lines denoting the vessel width measurement for branch points and black lines denoting the vessel width measurement for straight vessels.

(a) shows the vessel width measurement of Figure 1.

doi:10.1371/journal.pone.0049668.g005

We compared the performance of our algorithm with the performance of the two human experts. As mentioned, the branch center is defined as (x_k, y_k), where $k = 1, 2$ are the two experts. Similarly, we determine the branch center calculated in our algorithm as $(x_{alg}, y_{alg}) = (\frac{1}{2n}\sum_{i=1}^{n}(x_{1i} + x_{2i}), \frac{1}{2n}\sum_{i=1}^{n}(y_{1i} + y_{2i}))$, where n is the number of measurement profiles given by the algorithm. When the performance is compared, (x_1, y_1), (x_2, y_2), (x_{alg}, y_{alg}) are matched. A match is considered successful if the Euclidean distance between the two centers is less than 15 pixels (approximately one vessel diameter). The match could fail when (x_{alg}, y_{alg}) is too far from either (x_1, y_1) or (x_2, y_2), or (x_1, y_1) and (x_2, y_2) are too far from each other. The performance for a branching is considered as valid if the matches of all three branches are successful. Branchings were detected automatically. Among the 200 branchings, 3 out of 200 expert selected branchings were not detected by our automatic approach and were not included in further analysis. Hence, 197 out of 200 (98.5%) of the branches were used for further vessel width comparison. The vessel width is compared for valid branchings. The average of W_1 and W_2 is considered as the ground truth for comparison and is denoted as AVE in Table 1, which shows the signed errors and unsigned errors. The result of the comparison is given in Table 2. $branch_1$, $branch_2$ and $branch_3$ are the three branches.

Table 1: Comparison of the performance between human experts and presented method (signed error and unsigned error in pixels)

		E_1 vs E_2			Alg vs Ave		
		branch₁	branch₂	branch₃	branch₁	branch₂	branch₃
Signed Error	μ	-1.160	-1.440	-1.024	-0.408	-0.512	-0.102
	σ	1.587	1.662	-1.602	-1.679	-1.895	-2.090
Unsigned Error	μ	1.544	1.644	1.699	1.213	1.460	1.534
	σ	1.176	1.415	1.378	1.174	1.294	1.419

E_1 and E_2 denote the two human experts. Alg denotes the presented method. Ave denotes the average measurements of the two human experts. μ and σ are the mean and standard deviation of errors.
doi:10.1371/journal.pone.0049668.t001

doi:10.1371/journal.pone.0049668.t001

Table 2: Comparison of the performances between human experts and presented method (vessel width in pixels)

	branch₁		branch₂		branch₃	
	μ	σ	μ	σ	μ	σ
E_1	18.24	4.82	14.57	4.53	13.37	3.66
E_2	19.40	4.94	16.01	4.34	14.39	3.52
Alg	18.41	4.81	14.75	4.20	13.78	3.69

E_1 and E_2 denote the two human experts. Alg denotes the presented method. μ and σ are the mean and standard deviation of vessel width measurements.
doi:10.1371/journal.pone.0049668.t002

doi:10.1371/journal.pone.0049668.t002

The scatter plots are given in Figure 6. Arteries and veins were plotted separately. The Intraclass Correlation Coefficient is given to quantify how consistent it is between different measurements. The measurements for venous branches showed a better consistency both between experts, and between experts and the automatic algorithm.

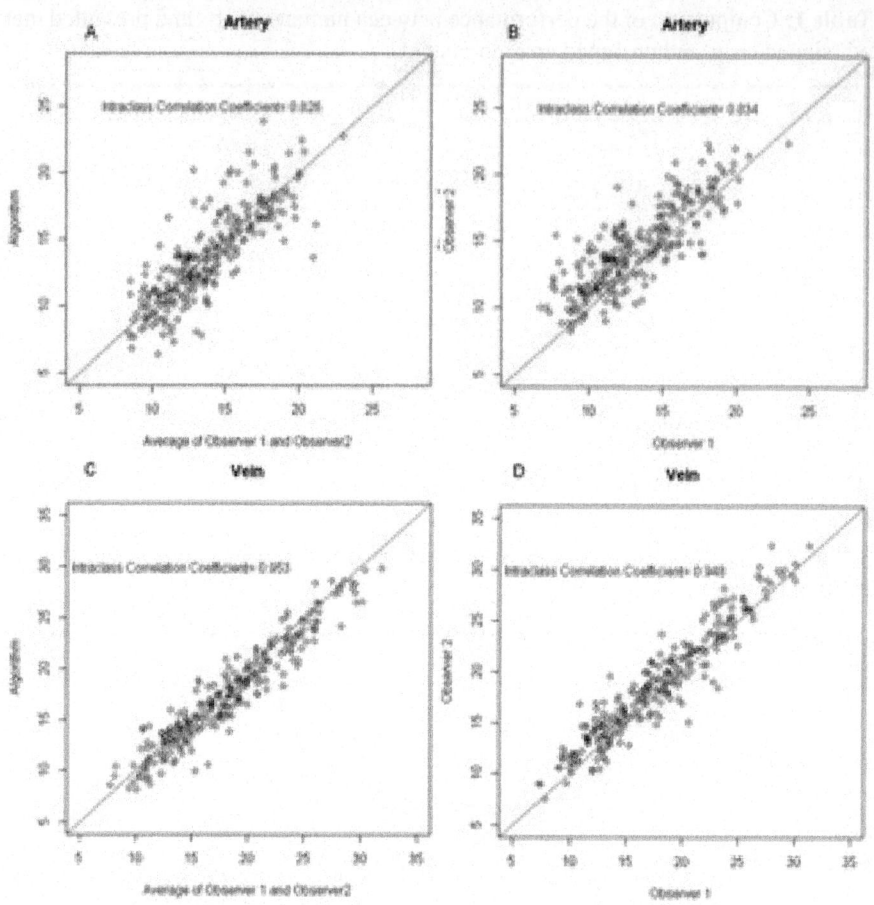

Figure 6: Vessel width measurement scatter plots in pixel.

(a) The scatter plot of vessel width measured by automatic algorithm and the average of vessel width given by expert 1 and expert 2 for artery branchings. (b) The scatter plot of vessel width measurement performed by expert 1 and expert 2 for artery branchings. (c) The scatter plot of vessel width measured by automatic algorithm and the average of vessel width given by expert 1 and expert 2 for vein branchings. (d) The scatter plot of vessel width measurement performed by expert 1 and expert 2 for vein branchings.

doi:10.1371/journal.pone.0049668.g006

The Bland-Altman plots are given in Figure 7.

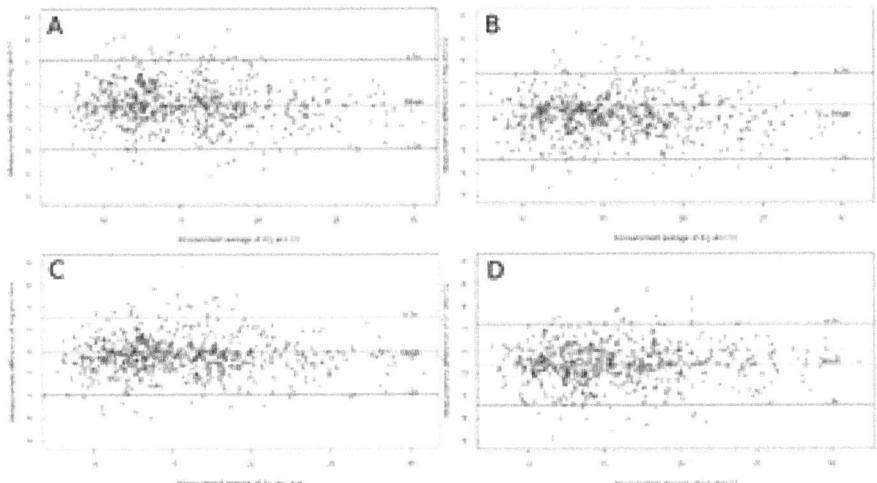

Figure 7: Bland-Altman plots, in pixel.

(a) Bland-Altman plot of proposed method and expert 1. (b) Bland-Altman plot of proposed method and expert 2. (c) Bland-Altman plot of proposed method and the average of expert 1 and expert 2. (d) Bland-Altman plot of expert 1 and expert 2.

doi:10.1371/journal.pone.0049668.g007

DISCUSSION

We developed and validated a fully automatic vessel width measurement method for branch points on retinal images. Performance of the method was comparable to the performance of human experts. Both the human experts and the automatic method showed a lower performance for artery compared to vein branch points. We observed that the veins usually have a more clearly defined vessel boundary than arteries, because of the increased absorption of light by de-oxygenated (venous) hemoglobin versus oxygenated (arterial) hemoglobin. For instance, Figure 8 shows a typical artery branch and a typical vein branch. This might have resulted in the ambiguities in arterial vessel width measurement for both human experts and the proposed automatic method.

Figure 8: Examples of typical arterial branch point and venous branch point.

Arteries usually have a lower contrast to the background comparing with veins. (a) A typical arterial branch point. (b) A typical venous branching.

doi:10.1371/journal.pone.0049668.g008

The proposed method is computationally efficient. For a retinal image of 2196×1958 pixels in size, the vessel segmentation map generation and image centerline generation takes less than 10 seconds. The average number of straight vessel segments on the image is 90 and the number of branching is about 30. It takes around 17 seconds to solve the graph for straight vessels and about 9 seconds to solve the graph for branch points.

The method suffers from the limitation that it relies largely on the quality of the vessel segmentation and centerline image. If the vessel centerline image exhibits a false positive branch point, the algorithm will still try to delineate the vessel boundaries, resulting in invalid measurements. However, our algorithm is not dependent on any specific vessel segmentation approach, and therefore, as improved vessel segmentations become available, our algorithm can be expected to improve as well.

An important application of the vessel width relationship at branch points is in artery-to-vein ratio (AVR). In the area of retinal image analysis, AVR is an important parameter, associated with stroke and other cardiovascular events in adults, and associated with an increased risk of retinopathy of prematurity in premature infants [7]. AVR is defined as the ratio between *Central Retinal Artery Equivalent* (CRAE) and *Central Retinal Vein Equivalent* (CRVE) [15], [16]. The vessel width relationship at arterial branch points and venous branch points is the most important information in AVR calculation. However, until now, only manual or semi-automatic methods are available to measure this relationship [1]–[4],[11],[16]. Due to the laboriousness of manual and semi-automatic work, only a limited number of measurements have been available to estimate the relationship, which reduced the reliability of the measurement. Future work of this study includes the application of the proposed method in AVR calculation.

CONCLUSIONS

We proposed a retinal vessel width measurement method at branch points based on an improved electric field theory motivated graph approach. This first fully automatic retinal branching vessel width measurement method has a performance comparable to human experts.

ACKNOWLEDGMENTS

The authors would like to thank Dr. Yin Yin for the helpful discussions.

AUTHOR CONTRIBUTIONS

Conceived and designed the experiments: MA XX. Performed the experiments: XX. Analyzed the data: XX. Contributed reagents/materials/analysis tools: BB PT GB. Wrote the paper: XX MA JR. Shared helpful suggestions during the implementation of method: QH.

REFERENCES

1. Doubal FN, de Haan R, MacGillivray TJ, Cohn-Hokke PE, Dhillon B, et al. (2010) Retinal arteriolar geometry is associated with cerebral white matter hyperintensities on magnetic resonance imaging. International Journal of Stroke 5: 434–439. doi: 10.1111/j.1747-4949.2010.00483.x

2. Parr JC, Spears GFS (1974) Mathematic Relationships Between the Width of A Retinal Artery and The Width of Its Branches. American Journal of Ophthalmology 77: 478–483.

3. Chapman N, Dell'omo G, Sartini MS, Witt N, Hughes A, et al. (2002) Peripheral vascular disease is associated with abnormal arteriolar diameter relationships at bifurcations in the human retina. Clinical Science 103: 111–116. doi: 10.1042/cs20010281

4. Stanton AV, Wasan B, Cerutti A, Ford S, Marsh R, et al. (1995) Vascular Network Changes in the Retina width Age and Hypertension. Journal of Hypertension 13: 1724–1728. doi: 10.1097/00004872-199512010-00039

5. Abràmoff MD, Garvin MK, Sonka M (2010) Retinal Imaging and Image

Analysis. IEEE Reviews in Biomedical Engineering 3: 169–208. doi: 10.1109/rbme.2010.2084567

6. Patton N, Aslam TM, Macgillivray T, Pattie A, Deary IJ, et al. (2005) Retinal vascular analysis as a potential screening tool for cerebrovascular disease: a rationale based on homology between cerebral and retinal microvasculatures. Journal of Anatomy 206: 319–348. doi: 10.1111/j.1469-7580.2005.00395.x

7. Sun C, Wang JJ, Mackey Da, Wong TY (2003) Retinal Vascular Caliber: Systemic, Environmental, and Genetic Associations. Survey Ophthalmol 48: 245–255. doi: 10.1016/j.survophthal.2008.10.003

8. Tsai CL, Stewart CV, Tanenbaum HL, Roysam B, Steward CV, et al. (2004) Model-based method for improving the accuracy and repeatability of estimating vascular bifurcations and crossovers from retinal fundus images. IEEE Transactions on Information Technology in BioMedicine 8: 122–130. doi: 10.1109/titb.2004.826733

9. Bhuiyan A, Nath B, Chua J, Ramamohanarao K (2008) Automatic detection of vascular bifurcations and crossovers from color retinal fundus images. In: International IEEE Conference on Signal-Image Technologies and Internet-Based System. IEEE, pp. 711–718.

10. Shen H, Roysam B, Steward CV, Turner JN, Tanenbaum HL, et al. (2001) Optimal scheduling of tracing computations for real-time vascular landmark extraction from retinal fundus images. IEEE Transactions on Information Technology in BioMedicine 5: 77–91. doi: 10.1109/4233.908405

11. Patton N, Aslam T, MacGillivray T, Dhillon B, Constable I (2006) Asymmetry of Retinal Arteriolar Branch Widths at Junctions Ability of Formulae to Predict Trunk Arteriolar Widths. Investigative Ophthalmology & Visual Science 47: 1329–1333. doi: 10.1167/iovs.05-1248

12. Quelhas P, Boyce J (2003) Vessel Segmentation and Branching Detection using an Adaptive Profile Kalman Filter in Retinal Blood Vessel Structure Analysis. In: Conference on Pattern Recognition and Image Analysis, ibPRIA. Springer-Verlag LNCS, pp. 802–809.

13. Calvo D, Ortega M, Penedo MG, Rouco J (2011) Automatic detection and characterisation of retinal vessel tree bifurcations and crossovers in eye fundus images. Computer methods and programs in biomedicine 103: 28–38. doi: 10.1016/j.cmpb.2010.06.002

14. Lee S, Reinhardt JM, Cattin PC, Abràmoff MD (2010) Objective and Expert-independent validation of retinal image registration algorithms

by a projective imaging distortion model. Medical Image Analysis 14: 539–549. doi: 10.1016/j.media.2010.04.001

15. Hubbard LD, Brothers RJ, King WN, Clegg LX, Klein R, et al. (1999) Methods for evaluation of retinal Microvascular abnormalities associated with hypertension/sclerosis in the atherosclerosis risk in communities study. Ophthalmology 106: 2269–2280. doi: 10.1016/s0161-6420(99)90525-0

16. Knudtson MD, Lee KE, Hubbard LD, Wong TY, Klein R, et al. (2003) Revised formulas for sum-marizing retinal vessel diameters. Current Eye Research 27: 143–149. doi: 10.1076/ceyr.27.3.143.16049

17. Chapman N, Witt N, Gao X, Bharath AA, Stanton AV, et al. (2001) Computer algorithms for the automated measurement of retinal arteriolar diameters. British Journal of Ophthalmology 85: 74–79. doi: 10.1136/bjo.85.1.74

18. Xu X, Niemeijer M, Song Q, Sonka M, Garvin MK, et al. (2011) Vessel boundary delineation on fundus images using graph-based approach. IEEE Transactions on Medical Imaging 30: 1184–1191. doi: 10.1109/tmi.2010.2103566

19. Niemeijer M, Staal J, van Ginneken B, Loog M, Abramoff MD (2004) Comparative study of Retinal Vessel Segmentation Methods on a New Publicly Available Database. In: SPIE Medical Imaging. Spie, volume 5370, pp. 648–656.

20. Sonka M, Hlavac V, Boyle R (1998) Image Processing, Analysis, and Machine Vision. New York: Thomson Learning, 3rd edition.

21. Yin Y, Song Q, Sonka M (2009) Electric Field theory motivated graph construction for optimal medical image segmentation. In: Graph Based Representations in Pattern Recognition. Springer, pp. 334–342.

22. Lee S, Abràmoff MD, Reinhardt JM (2010) Retinal atlas statistics from color fundus images. In: SPIE Medical Imaging. volume 7623.

23. Freeman WT, Adelson EH (1991) The Design and Use of Steerable Filters. IEEE Transactions on Pattern Analysis and Machine Intelligence 13: 891–906. doi: 10.1109/34.93808

24. Li K, Wu X, Chen DZ, Sonka M (2006) Optimal Surface Segmentation in Volumetric Images-A Graph Theoretic Approach. IEEE Transactions on Pattern Analysis and Machine Intelligence 28: 119–134. doi: 10.1109/tpami.2006.19

25. Jacobsen BK, Eggen AE, Mathiesen EB, Wilsgaard T, Njø lstad I (2011) Cohort profile: The Tromso Study. International Journal of Epidemiology : 1–7.

26. Christopher M, Moga DC, Russell SR, Folk JC, Scheetz T, et al. (2012) Validation of Tablet-Based Evaluation of Color Fundus Images. Retina 0: 1–7. doi: 10.1097/iae.0b013e3182483361

Chapter 11

ANTENNA MEASUREMENT

Dominique Picard

Supélec Plateau de Moulon 91192 Gif sur Yvette Cedex France

INTRODUCTION

The antenna is an important element of radiocommunication, remote sensing and radiolocalisation systems. The measurement of the antenna radiation pattern characteristics allows one to verify the conformity of the antenna. The simplest measurement method consists in the direct far-field measurement. For large antennas, the necessary measurement distance raises a problem which was resolved by the introduction of compact ranges and near-field techniques. Compact range consists in a focusing system, as a reflector, which can create a plane wave at short distance. The principle of near-field techniques is to measure the field radiated by an antenna at a short distance on a given surface surrounding the antenna, then to calculate the far-field starting from the measured near-field. These last techniques also make it possible to have an excellent precision as required by the satellite antennas for example. The near-field techniques also make it possible to carry out the diagnosis of the antennas, i.e. to find defects on the antenna. The duration of the measurement of the large antennas, which claims a large number of measurement points, was reduced considerably by the use of rapid near-field assessment system, for which the mechanical displacement of the probe was replaced by the electronic scanning of a probes array.

DIRECT FAR-FIELD MEASUREMENT

Antenna Pattern Measurement

There are four different regions for the electromagnetic field radiated by an antenna: three near-field regions and one far-field region. The nearest region is the reactive field region which extends until a distance of one wavelength l from the antenna surface. The second near-field region is the Rayleigh region which extends from the reactive region until a distance $D^2/(2l)$ from the antenna surface, a relation in which D is the tested antenna diameter. The third near-field region is the Fresnel region which extends from $D^2/(2l)$ until $2D^2/l$ from the antenna surface. The last region is the far-field Fraunhofer region which starts at a distance of $2D^2/l$. The space variations of the electromagnetic field differ in these four regions. It is thus necessary to be at a distance sufficient ($>2D^2/l$) to carry out direct far-field measurements.

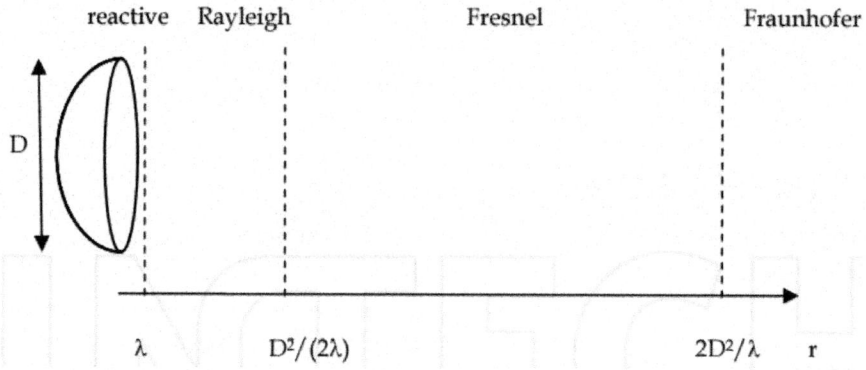

Figure 1: Different field regions for a large antenna.

Direct far-field measurement can be realized either in indoor or outdoor range (Kummer & Gillepsie, 1978). Indoor range consists in an anechoic chamber with one source antenna and the tested antenna placed on a positioner. This positioner allows one to vary the tested antenna attitude with respect to the direction of the wave incidence of the wave for far-field pattern measurement. Outdoor range consists in a tower bearing a source antenna and the tested antenna placed on a positioner. The distance between the two antennas can be larger for outdoor range, i.e. the capacity of outdoor range in terms of tested antenna dimensions is higher. On the other hand the outdoor range is sensitive to parasitic signals and ground reflections.

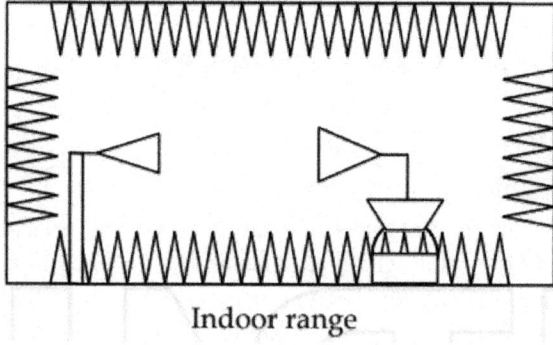

Indoor range

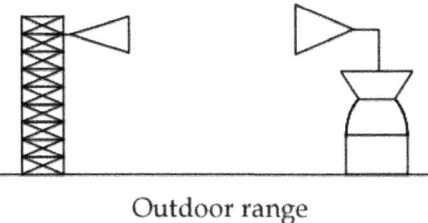

Outdoor range

Figure 2: Different range geometries.

Measurement of the Characteristics of the Pattern

Gain Measurement

Absolute gain measurements use power budget between antennas. The Friis transmission formula gives a relation between the gains G_A and G_B of two antennas A and B:

$$P_r = P_i \ G_A \ G_B \ (\lambda/4\pi r)^2$$

(1)

P_r is the power received at a matched load connected to the receiving antenna, P_i the power accepted by the transmitting antenna, l is the wavelength and r the distance between the two antennas. The use of this formula requires that the two antennas A and B are polarization matched and that the separation distance between the antennas corresponds to far-field conditions.

Transmitting antenna   Receiving antenna

Figure 3: The two antennas system corresponding to the Friis formula.

The first method uses the tested antenna and a reference antenna with a known gain. The measurements of P_r, P_i, r, and the knowledge of l give the value of the tested antenna gain by means of the Friis formula. The second method used two identical tested antennas with the same gain. The Friis formula gives the value of the tested gain antenna starting from the same measurements as the preceding method. The third method used three antennas A, B and C, generally the tested antenna and two other antennas. Three different power budgets are carried out, with the three possible antennas pairs leading to the three following equations:

$$G_A \, G_B = (P_r/P_i)_{AB} \; (4\pi r/\lambda)^2 \tag{2}$$

$$G_B \, G_C = (P_r/P_i)_{BC} \; (4\pi r/\lambda)^2 \tag{3}$$

$$G_C \, G_A = (P_r/P_i)_{CA} \; (4\pi r/\lambda)^2 \tag{4}$$

These three formulas provide the determination of the three different gains G_A, G_B and G_C.

Directivity Measurement

There are two methods which make it possible to measure the directivity Dir of an antenna. The first method is based on the definition of the directivity. The knowledge of the relative far-field radiated by the antenna in all the directions is sufficient to know the directivity of this antenna. This method is called the pattern integration method because it uses the integration of the radiated power density (dPe/dS) on all the directions to calculate the total power radiated by the antenna Pe. The power density is related to the electric field E at a distance r by the relation:

$$(dP_e/dS) = E^2/Z_0 \tag{5}$$

$Z_0 = 120p$ W is the free space impedance of wave.

$$P_e = \int_{\varphi=0}^{2\pi} \int_{\theta=0}^{\pi} (dP_e/dS) \; r \sin\theta \; d\phi \; d\theta \tag{6}$$

$$Dir = (dP_e/dS) \, / \, [P_e/(4\pi r^2)] \tag{7}$$

The second method uses the relation between the directivity Dir and the gain G of a given antenna:

$$Dir = G/\eta \tag{8}$$

in which h is the efficiency of the antenna. The measurements of the gain and the efficiency of the antenna result in the knowledge of the directivity.

The first method is used more for directive antennas while the second is rather used more for omnidirectional antennas.

Efficiency

There are essentially three different methods to measure the efficiency of an antenna. The first method consists in the measurement of the gain G and the directivity D of the antenna, and then the relation between gain, directivity and efficiency h results in to obtain the efficiency:

$$\eta = G/Dir \tag{9}$$

The second method uses the measurement of the antenna input impedance and is called the Wheeler Cap Method. For certain antennas, microstrip patches for example, the losses can be modeled by a series resistance or a parallel resistance R_l with the antenna input resistance R_r. Two different antenna input impedance measurements are performed. The first measurement is done with the antenna in free space and the second measurement with the antenna inside a metallic hemisphere, and the input resistance of the antenna takes the values R_1 and R_2 respectively for these two measurements. If the loss resistance occurs in series, then it is short-circuited by the cap, and if the loss resistance occurs in parallel, then it is open-circuited by the cap. It is possible to have the relation between the efficiency, the radiated power and the losses power. This relation allows one to obtain the efficiency in function of R_r and R_l and then in function of the measured input resistance R_1 and R_2. For the series modelling the efficiency is:

$$\eta = R_r/(R_r + R_l) = (R_1 - R_2)/R_1 \tag{10}$$

And for the parallel modelling:

$$\eta = R_l/(R_r + R_l) = (R_2 - R_1)/R_2 \tag{11}$$

The third method is called radiometric method. It consists in the measurement of the available noise power at the output of the antenna by means of a radiometer (Ashkenazy et al, 1985). This power U is related to the effective temperature T_e of the antenna and the equivalent temperature T_n of the radiometer by:

$$U = C(T_e + T_n) \tag{12}$$

in which C is a constant. The equivalent temperature T_n is derivated from the noise figure of the radiometer. The effective temperature T_e of the antenna, at physical temperature Ta, is related to the efficiency h of the antenna by:

$$T_e = T_a (1-\eta) + T_t \eta \tag{13}$$

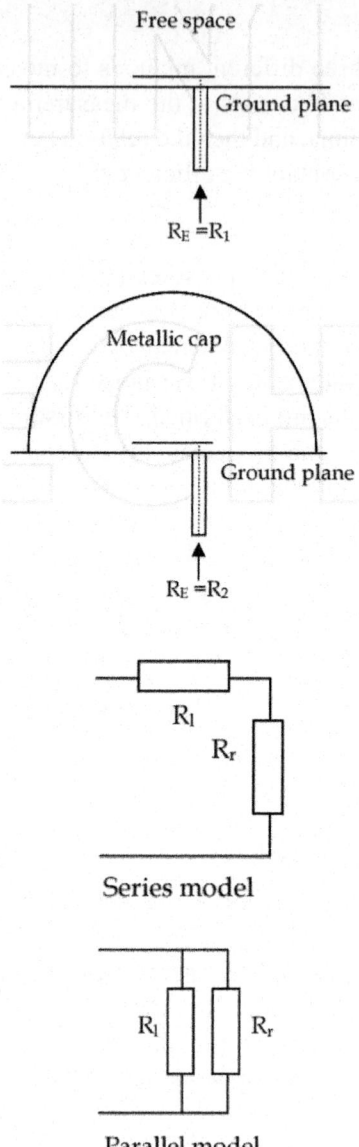

Figure 4: The Wheeler Cap Method.

T_t is the temperature target aimed by the antenna. The measurement of the noise power U for two different target temperatures, respectively the cold temperature T_c and the warm temperature T_w, allows the determination of the efficiency h. The cold target temperature is a clear sky and the warm one is an extended absorber at room temperature T_a.

$$\eta = [(T_n + T_a)(1 - \varepsilon)]/ [\varepsilon (T_c + T_a)] \tag{14}$$

$$\varepsilon = U_c/U_w \tag{15}$$

In fact it is also better to measure another antenna with high efficiency, such as a horn, for which the efficiency is equal to 1, with the same cold and warm temperatures as the first antenna. For this second antenna:

$$U_c/U_w = \delta = (T_n + T_a)/ (T_n + T_c) \tag{16}$$

The efficiency h of the first antenna is:

$$\eta = [\delta (\varepsilon - 1)]/[\varepsilon (\delta - 1)] \tag{17}$$

Clear sky

U_c

Absorber material
Tempearture Ta

U_w

Figure 5: The radiometric method.

To achieve high sensitivity measurements the radiometer should have a low internal noise temperature and the temperature difference between the two targets should be as high as possible. The gain and directivity method and the radiometric method are well suited for directive antennas and the Wheeler Cap method for small antennas. A comparison study of these three methods for microstrip antennas has an accuracy of about 2% for the Wheeler Cap method, 10% for the radiometric method and 20% for the gain and directivity method (Pozar & Kaufman, 1988).

Polarization Measurement

For harmonic mode, the electromagnetic field radiated by an antenna is polarized. Generally speaking, in one time period, the electric and magnetic fields, observed at a given point, describe a plane curve which is an ellipse. In the case of linear polarization the ellipse is reduced to its major axis. For circular polarization the modulus of the field remains constant. The knowledge of the polarization is equivalent to the knowledge of the ellipse: its axial ratio, the slope of its major axis relative to a reference direction and the sense of the displacement along the ellipse.

The experimental determination of the characteristics of the polarization ellipse can be carried out in several ways. It is possible to use amplitude-only measurements, or two amplitude and phase measurements with two different antennas with independent polarizations. The simplest method consists in using two antennas with linear polarization orthogonal one with the other. The same antenna can be used with two different orientations with orthogonal linear polarizations. Measurements provide the amplitude and the phase of the two field orthogonal components from which it is possible to calculate the characteristics of the polarization ellipse.

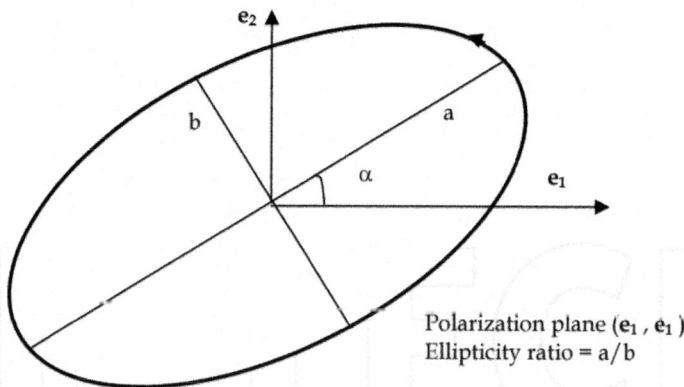

Figure 6: Characteristics of the polarization ellipse.

COMPACT ANTENNA TEST RANGES

Classical Compact Antenna Test Ranges

The compact ranges allow the direct far-field pattern measurement of antennas at near-field distance. It is then possible to carry out indoor antenna far field pattern measurements. The principle of compact range is to produce a plane wave at short distance, by means of a focusing system, like a reflector or a lens, illuminated by a primary feed (Johnson et al, 1969). There are two principal kinds of compact ranges. The first kind uses an offset parabolic reflector, with a possible Cassegrain geometry. The second kind uses two orthogonal cylindrical parabolic reflectors. This second geometry is equivalent to the first one with a significantly reduced cost due to the simpler reflectors geometry. The choice of an offset feed prevents aperture blockage and reduces the diffracted energy from the feed structure in the test region.

The main causes of errors on the field in the quiet zone are direct radiation from the primary feed, diffraction from the feed support, diffraction from the edges of the reflector, reflector surface deviations from the theoretical surface and room reflections. For surface deviations two parameters have to be taken into account: the value of the deviation and its extent over the surface. Calculations (Johnson et al, 1973) show that a 0.5dB error on the field corresponds to a 0.007 wavelength surface deviation. The use of serrations or rolled edges reduces the effect of the diffraction from the reflector edges. The positions and the length of the serrations are empirically adjusted so that the energy is diffracted in directions away from the test region for a broad band of frequencies. Room reflections and diffraction from the feed support are minimized by the use of absorbing material. The wise choice of the primary feed position compared to the tested antenna reduces the coupling due to the direct radiation of the feed. Time-gating of the measurement signal improve the performances of the compact range in relation to these last three errors causes.

The dimensions of the quiet zone are about the third of the reflector dimensions. Broadband compact ranges are available: 0.7GHz-100GHz with 3.6m cubic quiet zone. For such ranges, several feed antennas are used, each antenna covering a half octave frequency band. The different feed antennas are automatically positioned at the reflector focus point and connected to the instrumentation, according to the required frequency. Typical amplitude variations of 1dB and phase variations of 10°, in the quiet zone, can be achieved. The cross polarization is better than 30dB.

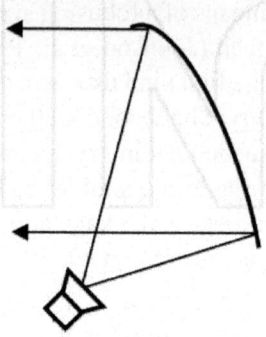

Parabolic reflector

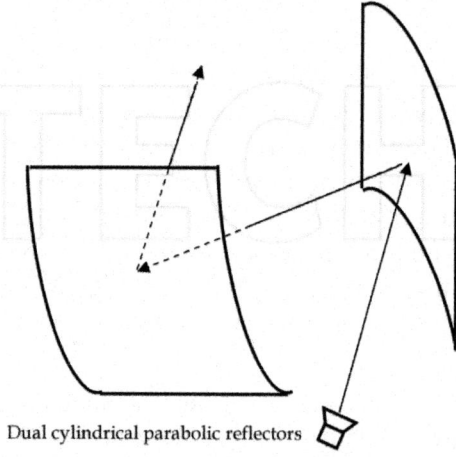

Dual cylindrical parabolic reflectors

Figure 7: Different kinds of compact ranges.

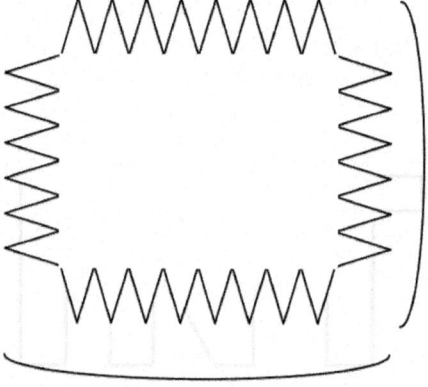

Figure 8: Compact range reflector with serrations or rolled edges.

Hologram Compact Antenna Test Range

It is difficult to test large antennas at frequencies above 100GHz. The use of near-field antenna measurements requires deformable coaxial cables or rotary joints with high performances, which is impossible at frequencies that are too high. Conventional reflector type compact antenna test ranges require one or more reflectors for which the surface accuracy needs to be better than about 0.01 wavelength, that is to say, for example, 15µm at 200GHz.

The use of a planar hologram constitutes another solution (Hirvonen et al, 1997). The surface accuracy requirements for an amplitude hologram are less stringent than those for a reflector, and its planar geometry simplifies its realization. The hologram compact antenna test range is a low-cost and easy-to-fabricate structure. The principle of the hologram is to change the spherical wave front radiated by a source antenna (a horn for example) in a plane wave by means of the transmission through the hologram. It is possible to numerically calculate the structure required to change the known input field into the desired output field. The fabrication of the hologram is simplified by binary amplitude quantization: the local transmittance of the hologram is either 1 or 0. This is obtained by the use of a copper-plated Kapton film using an etching procedure. A hologram of 3 meters diameter has been used for the test of a 1.5 meter diameter reflector antenna at 322GHz. The envelopes of the measured and simulated far-field patterns are similar, but there are relatively important differences between the two far-field patterns.

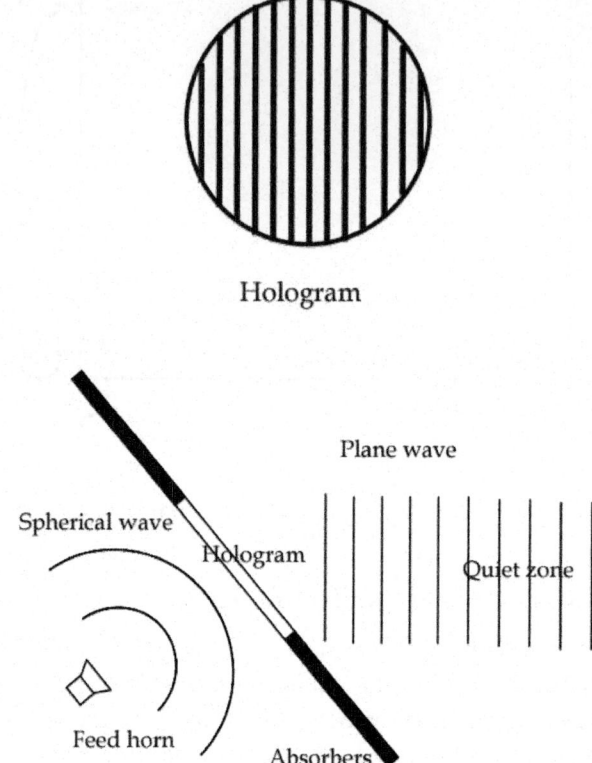

Hologram

Figure 9: Hologram and hologram compact antenna test range.

NEAR-FIELD TECHNIQUES AND APPLICATIONS

Near-Field Techniques

The principle of near-field techniques is to measure the field radiated by an antenna at a short distance on a given surface surrounding the antenna, then to calculate the far-field starting from the measured near-field (Yaghjian, 1986). Of the several formulations for these techniques, the two principal formulations are the Huygens principle and the modal expansion of the field.

Huygens Principle

The tangential components of the electric and magnetic fields E_t and H_t are measured on an arbitrary surface S enclosing the tested antenna. These components allow one to calculate the equivalent electric and magnetic currents

J_s and M_s. Then the electric and magnetic fields can be evaluated everywhere out of the surface S starting from the equivalent currents. This method uses simple calculations, but for large antenna of diameter D the computer time varies like $(D/l)^3$ and can become very long. Moreover the method requires calibrated and ideal probes and generally the measurement of the four field components. The electric and magnetic far-field E and H are given by the relations:

$$J_s = n \times H_t \qquad M_s = -n \times E_t \tag{18}$$

$$E = -j\, k/(4\pi) \int\int_S [Z_0\, (J_s \times u) \times u - M_s \times u]\, e^{-jkr}/r\, dS \tag{19}$$

$$H = -j\, k/(4\pi) \int\int_S [J_s \times u + 1/Z_0\, (M_s \times u) \times u]\, e^{-jkr}/r\, dS \tag{20}$$

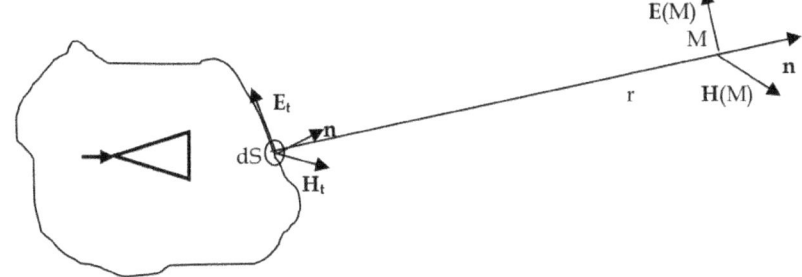

Figure 10: The Huygens principle.

Modal expansion of the field In free space the electric and magnetic fields verify the propagation equation. This equation has elementary solutions or modes and a given field is a linear combination of these modes. The knowledge of the field of an antenna is equivalent to the knowledge of the coefficients of the linear combination. The expression of the modes is known for the different systems of orthogonal coordinates: cartesian, cylindrical and spherical. The coefficients of the linear combination are obtained by means of the two tangential field components measurement on a reference surface of the used coordinates system, then using an orthogonality integration. The case of the planar scanning is simple (Slater, 1991). The measurement of the two tangential components of the field, the electric field $E_t(x,y,z)$ for example, is realized on a plane z=0 following a two dimensional regular grid (axis x and y). The antenna is located at z< 0. The tangential components of the plane wave spectrum are obtained from the measured field of the orthogonality integration:

$$A_t(k_x, k_y, z) = 1/(2\pi) \int_{-\infty}^{+\infty} \int_{-\infty}^{+\infty} E_t(x,y,z) \, e^{j(k_x x + k_y y)} \, dx \, dy$$

(21)

It is then possible to calculate the electric field in any point thanks to:

$$E(x,y,z) = 1/(2\pi) \int_{-\infty}^{+\infty} \int_{-\infty}^{+\infty} A(k_x, k_y) \, e^{-j(k_x x + k_y y + k_z z)} \, dk_x \, dk_y$$

(22)

$$k^2 = \omega^2 \varepsilon_0 \mu_0 \qquad\qquad k^2 = k_x^2 + k_y^2 + k_z^2$$

(23)

The normal component $A_z(k_x, k_y)$ of vector $A(k_x, k_y)$ is obtained from the local Gauss equation:

$$\mathbf{k} \, A(k_x, k_y) = 0 \qquad\qquad \mathbf{k} = k_x \, e_x + k_y \, e_y + k_z \, e_z$$

(24)

It is then possible to obtain the near-field of the antenna everywhere from the measurement of the near-field on a given plane. The electric far-field in the direction q,f and at a distance r is given by the relation:

$$E(r,\theta,\phi) = j \, k \, \cos\theta \, e^{-jkr}/r \, A(k\sin\theta\cos\phi, k\sin\theta\sin\phi) \qquad k^2 = \omega^2 \varepsilon_0 \mu_0$$ (25)

It would be possible to obtain the magnetic field from the Maxwell-Faraday equation with the knowledge of the electric field. The sampling spacing on the measurement surface is l/2 following rectilinear axis (planar and cylindrical scanning) and l/2(R+l) for angular variable (cylindrical and spherical scanning), R is the radius of the minimal sphere, i.e. the sphere whose centre is on the rotation axis, which contains the whole of the antenna and whose radius is minimal.

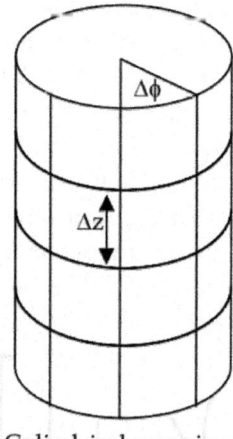

Cylindrical scanning

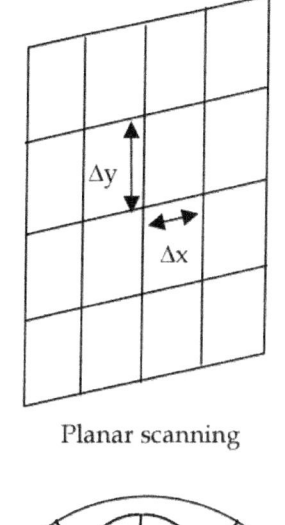

Planar scanning

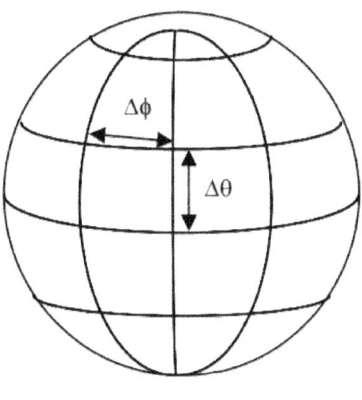

Spherical scanning

Figure 11: Sampling spacing for the different scanning geometries: $\Delta x = \Delta y = \Delta z = 1/2$, $\Delta f = \Delta q = 1/2(R+1)$

Probe Correction

In practice, the probe is not an ideal electric or magnetic dipole which measures the nearfield in a point. The far-field pattern of the probe differs appreciably from the far-field of an elementary electric and magnetic dipole. For the accurate determination of electric and magnetic fields from near-field measurements, it is necessary to correct the nonideal receiving response of the probe. The probe remains oriented in the same direction with planar scanning, and the sidelobe field is sampled at an angle off the boresight direction of the probe. Thus it is necessary to apply probe correction to planar near-field measurements. The problem is the same with cylindrical scanning for the rectilinear axis, and probe correction is also necessary in this case. For spherical scanning, the probe always points toward the test antenna and probe correction is not necessary if the measurement radius is large enough.

The formulation of probe correction is simple for planar scanning. The plane wave spectrum of the measurement A_m, as definite previously, is the scalar product of the plane wave spectra of the tested antenna A_a and the probe A_p:

$$A_m = A_a A_p = A_{ax} A_{px} + A_{ay} A_{py}$$

(26)

The measurement is repeated twice, for two orthogonal orientations between them, of the probe. This results in two equations on A_{ax} and A_{ay} and it is enough to invert this linear system of equations to obtain Aax and A_{ay}.

Different Coordinates Systems Comparison

In the case of planar cartesian and cylindrical coordinates systems, the measurement surface is truncated because the length of a rectilinear axis is limited. In practice, the measurement surface is a rectangle for planar exploration and a cylinder with a finite height for cylindrical exploration. Thus to minimize the effect of the measurement surface truncation, planar nearfield systems are devoted to two-dimensional directive antennas and the cylindrical system requires antennas with directive pattern in at least one plane. Spherical near-field systems are convenient for omni-directional and directive antennas.

Phaseless Method

The use of near-field techniques at frequencies above 100GHz is very difficult. This is due to the phase errors induced by coaxial cables or rotary joints whose performances are degraded at these frequencies. In counterpart, it is possible to measure the amplitude of the near-field until very high frequencies. This is why the phaseless methods appeared. These methods consist in the measurement of the near-field on two different surfaces, two parallel planes in front of the antenna for example, and to try to find the phase using an iterative process (Isernia & Leone, 1994). This iterative process consists in passing alternatively from one surface to the other by a near-field to near-field transformation. At the beginning, the distribution of the near-field phase on a surface is arbitrarily selected, a constant phase for example. Then when the near-field is calculated on the other surface, the calculated phase is preserved, and one associates it with the measured near-field amplitude. Then the near-field is calculated on the first surface and one starts again the process again. The process is stopped when the difference between the amplitudes of the computed and measured fields is lower than a given value.

To obtain an accurate reconstructed phase, it is necessary that the near-fields on the two planes are sufficiently different, i.e. the two planes are separated by a sufficient distance. A study shows good results for a low sidelobe shaped reflector antenna with an elliptical aperture with axes 155cm x 52cm at 9GHz (Isernia & Leone, 1995). The two planes are at a distance respectively of 4.2cm and 17.7cm from the antenna. The far-field pattern obtained from the near to far-field transformation with phaseless method shows agreement with the reference far-field pattern, up to a -25dB level approximately.

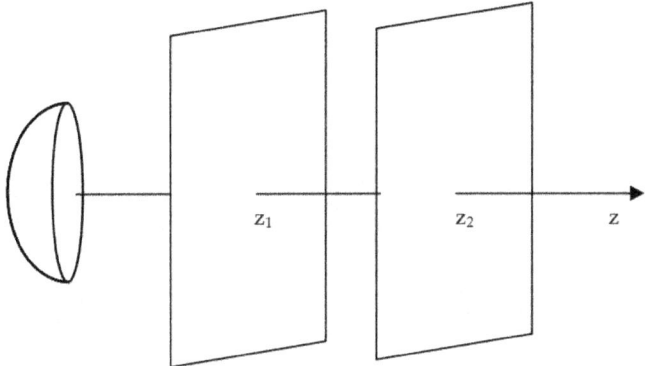

Figure 16: Phaseless method with two parallel planes configuration.

Near-Field Measurement Errors Analysis

One of the difficulties related to the use of the near-field techniques is the evaluation of the effect on the far field, of the measurement errors intervening on the near-field. A study allows the identification of the error sources, an evaluation of their level and the value of the induced uncertainties on the far-field, in the case of planar near-field measurements (Newell, 1988). About twenty different error sources are identified as probe relative pattern, gain, polarization, or multiple reflections between probe and tested antenna, measurement area truncation, temperature drift... The main error sources on the maximum gains are the multiple reflections between probe and tested antenna, and the power measurement, for a global induced error of 0.23dB. For sidelobe measurement, the main error sources are the multiple reflections between probe and tested antenna, the phase errors, the probe position errors and the probe alignment for a global induced error of 0.53dB on a -30dB sidelobe level. A comparison of the results obtained with four different near-field European ranges shows agreement on the copolar far-field pattern and directivity of a contoured beam antenna (Lemanczyk, 1988).

Near Field Applications

Electromagnetic Antenna Diagnosis

Antenna diagnosis consists in the detection of defects on an antenna. There are essentially two different electromagnetic diagnosis: reflector antenna diagnosis and array antenna diagnosis.

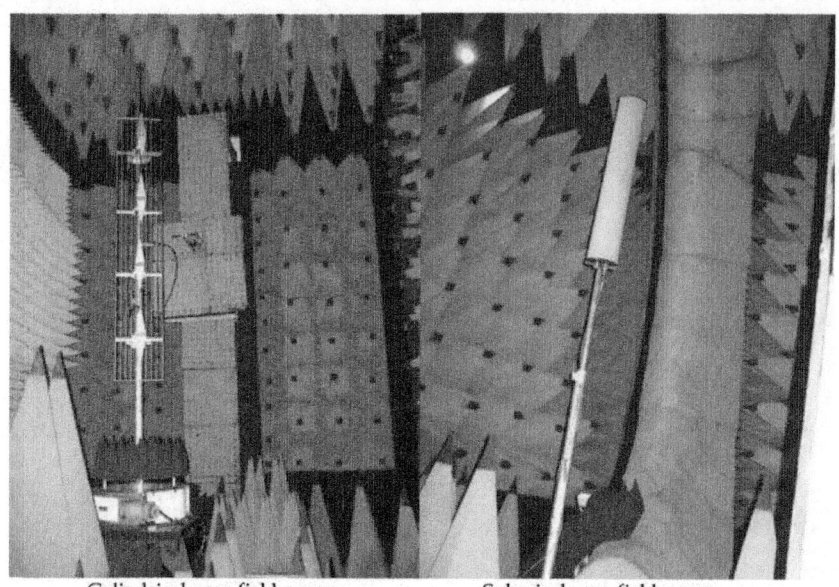

Cylindrical near-field range   Spherical near-field range

Figure 12: Near-field ranges at Supélec.

Reflector Antenna Diagnosis

For reflector antennas, the diagnosis consists mainly in checking the reflector surface. It is possible to use an optical method to measure the reflector surface. This is a photogrammetric triangulation method (Kenefick, 1971). This method utilizes two or more long-focal length cameras that take overlapping photographs of the surface. This surface is uniformly covered with self-adhesive photographic targets whose images appear on the photographic record. The two-dimensional measurements of the image of the targets are processed with a least squares triangulation to provide the three-dimensional coordinates of each target. The accuracy of this method is of the order of one part in 100000 of the reflector diameter.

It is also possible to perform electromagnetic diagnosis of reflector antenna (Rahmat Samii, 1985). For this method, the knowledge of the amplitude and phase far-field pattern is required. This far-field can be obtained by means of near-field, compact range or direct farfield measurement. The relation between the two-dimensional amplitude and phase far-field and the electric current on the reflector surface is known. This relation can take the form of a two-dimensional Fourier transform at the cost of some approximations, and can then be inverted easily. Finally, the phase of the currents can be interpreted like a deformation starting from the theoretical geometry of the reflector. A study of this method using spherical near-field measurements on a large reflector antenna give good results: small deformations of about one l diameter and a l/10 thickness are detected (Rahmat Samii, 1988).

Array Antenna Diagnosis

The electromagnetic diagnosis of array antennas consists in detecting defective or badly fed elements on the antenna. To obtain this detection, it is sufficient to rebuild the feeding law of the antenna elements. There are two methods of array antenna diagnosis that primarily exist. The first method uses backward transform from the measurement plane to the antenna surface and is called the spectral method (Lee et al, 1988). The measurement of the radiated near field is performed on a plane parallel to the antenna surface. Then the measured near field is processed to obtain the near field at the location of each element of the array. This processing contains element and probe patterns correction. The feeding of each element is then considered as being proportional to the near field at the location of the element. The second method uses the linear relation between the feeding of each element and the measured near field and is called the matrix method (Wegrowicz & Pokuls, 1991), (Picard et al, 1996), (Picard et al, 1998). The near field is also measured on a plane parallel to the antenna surface. The number of space points is higher than or equal to the number of elements in the array. The linear equation system is numerically inverted. The advantage of the matrix method, compared to the spectral method, is that it uses a number of measurement points significantly weaker. The accuracy of these methods on the reconstructed feeding law is of the order of a few degrees and a few tenth of dB.

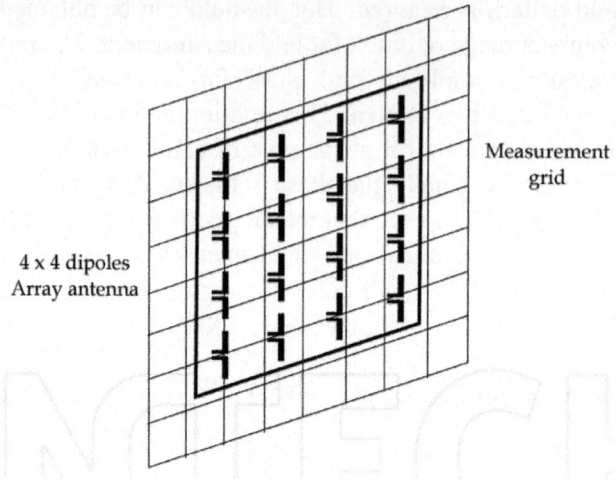

Figure 13: Array antenna diagnosis: measurement configuration.

Antennas Coupling

The coupling coefficient between two antennas can be obtained by using the fields radiated by these two antennas separately (Yaghjan, 1982). The reciprocity theorem makes it possible to show that the voltage V_{BA} induced by the radiation of an antenna A at the output of an antenna B is

$$V_{BA} = - \iint_S [E_a x H_b + H_a x E_b] \, n \, dS$$

(27)

S is a close surface surrounding the antenna B,

n is the normal vector to S with the outside orientation,

E_a, H_a electric and magnetic fields radiated by the antenna A,

E_b, H_b electric and magnetic fields radiated by the antenna A for the emission mode with unit input current,

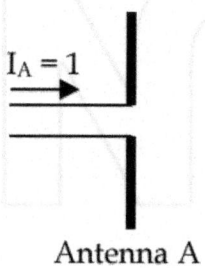

$I_A = 1$

Antenna A

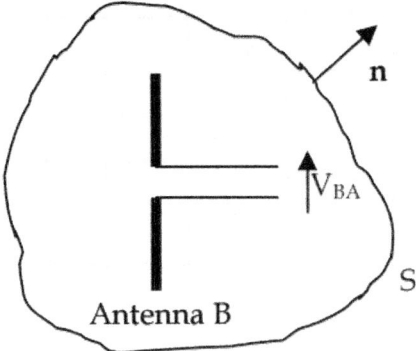

Figure 14: The two antennas system for coupling evaluation.

The advantage of this method is that it can predict, by calculation, the coupling between the two antennas for any relative position, only by means of their separate radiated near-fields measurements.

Determination of the Safety Perimeter of Base Station Antennas

An application of the cylindrical near-field to near-field transformations is the determination of the base station antennas safety perimeter. The electric and magnetic nearfields level can be evaluated from the near-field measurements and from the power accepted by the antenna. The comparison of this level with the ICNIRP reference level allows the determination of the safety perimeter (Ziyyat et al, 2001), (ICNIRP, 1998). The accuracy obtained by this method is within a few percent on the calculated near-field.

Rapid Near-Field Assessment System

The near-field measurement of a large antenna requires a considerable number of measurement points. Computers' computing power has increased regularly and was multiplied by approximately 100000 between 1981 and 2006. The result is from it that the duration of the far-field calculation decreases regularly and is no longer a problem. On the other hand the duration of measurement can be very important. This is due to the slowness of mechanical displacements. The replacement of the mechanical displacement of the probe by the electronic scanning of a probes array makes it possible to accelerate considerably the measurement rate and to reduce the measurement duration (Picard et al., 1992), (Picard et al, 1998).

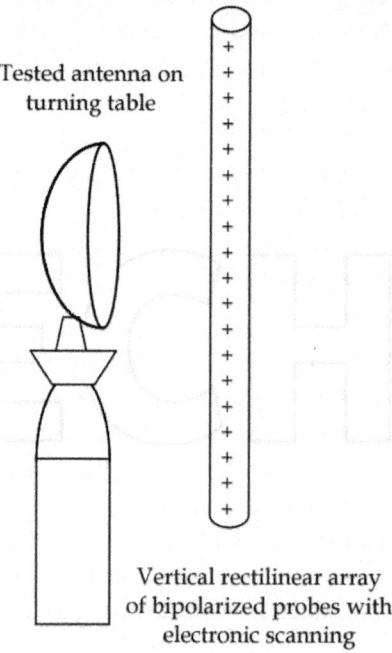

Tested antenna on
turning table

Vertical rectilinear array
of bipolarized probes with
electronic scanning

Figure 15: Rapid near-field range at Supélec and principle of rapid near-field assessment systems.

ELECTROMAGNETIC FIELD MEASUREMENT METHOD

The measurement of the radiation of the antennas is indissociable from the measurement of high frequency electromagnetic field. Primarily four different methods for high frequency electromagnetic field measurement exist. These methods differ primarily by the type of connection between the probe and the receiver, this connection could possibly be the cause of many disturbances. The first method is the simplest one. It consists in using of a small dipole probe connected to a receiver with a coaxial line. In order to limit the parasitic effects of the line on the measurement signal, a balun is placed between the line and the dipole. This method makes it possible the measurement of the local value of one component of the electric or magnetic field.

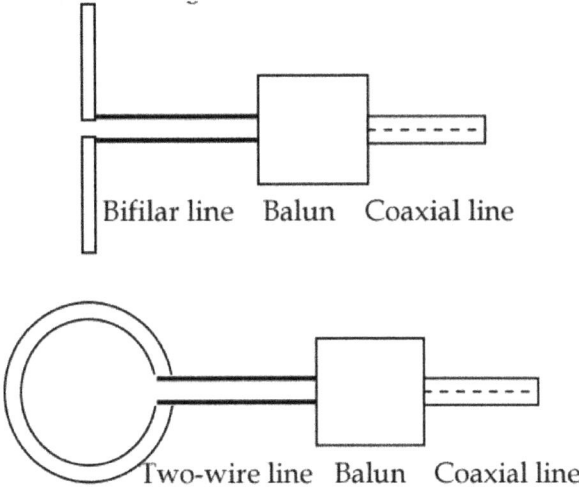

Figure 17: Measurement of the electric and magnetic field with a dipole probe.

In the case of a field whose space variations are very fast, the modulated scattering technique can be used advantageously. This second method consists in the use of a small probe loaded with a nonlinear element like a PIN diode, which is low frequency modulated (Callen & Parr, 1955), (Richmond, 1955), (Bolomey & Gardiol, 2001). The electromagnetic field scattered by this probe is collected by the emitting antenna (monostatic arrangement) or by a specific or auxiliary antenna (bistatic arrangement) called auxiliary antenna. The signal provided by the emitting antenna is proportional to the square of the field radiated at the probe location for the monostatic arrangement, and that provided by the auxiliary antenna is proportional to this field for the bistatic arrangement. These two signals are low frequency modulated like the scattered field, and this amplitude modulation allows one to retrieve this signal among parasitic signals, with coherent detection for example. The low frequency modulation of the diode may be conveyed by resistive lines or by an optical fiber in the case of the optically modulated scattering technique (Hygate & Nye, 1990) so as to limit the perturbations.

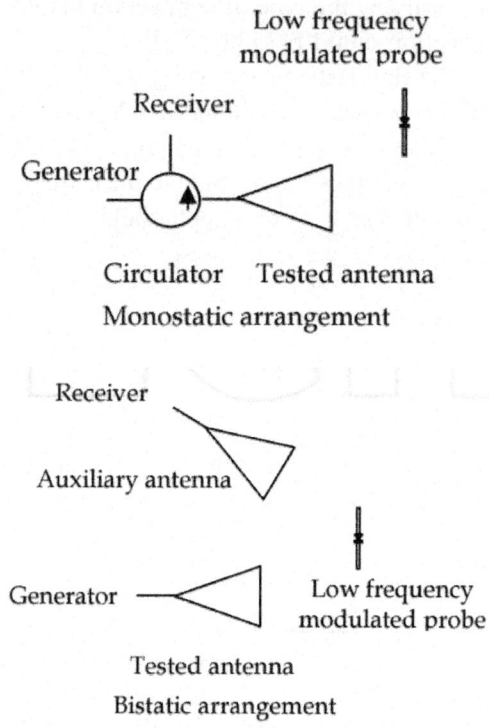

Figure 18: The modulated scattering technique.

The third method uses an electro-optic probe. This probe is a small one like a dipole, and is loaded with an electro-optic crystal like $LiNbO_3$. The refraction index of the crystal linearly depends on the radiofrequency electric field which is applied to it. The light of a laser is conveyed by an optical fiber and crosses the crystal. The phase variations of the light transmitted through the crystal are measured and are connected linearly to the radiofrequency electric field applied to the crystal. The calibration of the probe makes it possible to know the proportionality factor between the variation of the phase undergone by the light and the amplitude of the measured radiofrequency electric field. This method makes it possible to produce probes with very broad band performances (Loader et al, 2003). In particular, an electric dipole of this type is an excellent time-domain probe: the measurement signal is proportional to the measured time-domain electric field.

The last method is simpler and less expensive than the two preceding ones while making it possible to carry out very local measurements without the disturbances due to the connection between the probe and the receiver. This method uses detected probes (Bowman, 1973) to measure the local electric field. Such a probe is loaded with a schottky diode and detects the RF currents induced by the electric field, to obtain a continuous voltage. This voltage can be measured by a voltmeter. The lines connecting the dipole and the voltmeter are made highly resistive to reduce their parasitic effect. The main defects of this method are its poor sensitivity and that it provides only the amplitude of the measured field. If the knowledge of the phase is necessary for the application, it must be obtained by means of phaseless methods.

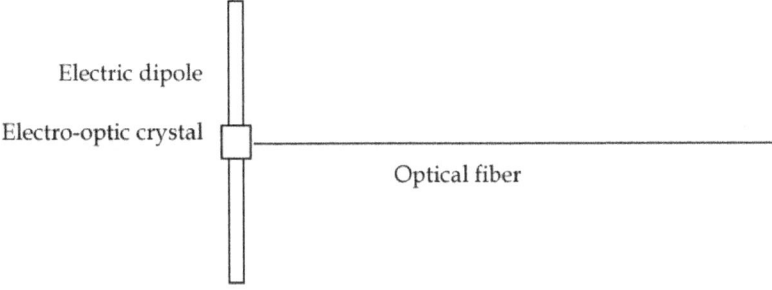

Figure 19: Electro-optic dipole probe.

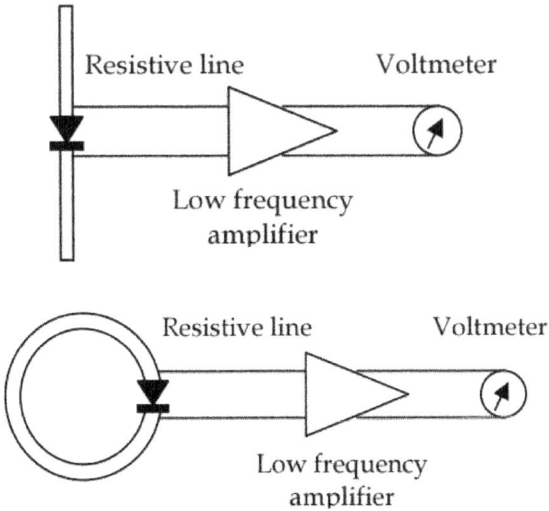

Figure 20: Detected probe.

INSTRUMENTATION

The instrumentation used for antenna measurements depends on the temporal mode used: time domain or frequency domain. Network or spectrum analyzer and frequency synthetizer are used for frequency domain measurements. Real time or sampling oscilloscope and pulse generator are used for time domain measurements.

Frequency Domain Antenna Measurements

The system of emission-reception the most used for antenna measurements is the vector network analyzer. It allows the measurement of transmission coefficients and it supplies the phase. Its intermediary frequency bandwidth can reach 1MHz, i.e. it allows very high speed measurements, and its dynamic can reach 140dB. It can have several ways of measurement so as to be able to measure simultaneously direct and cross polarizations. Its maximum frequency bandwidth of operation is 30kHz to 1000GHz (with several models). It is also possible to use a scalar network analyzer or a spectrum analyzer coupled to a frequency synthetizer when the measurement of the phase is not necessary as for far-field for example.

Time Domain Antenna Measurements

Antenna measurements in the time domain are less frequent than in the frequency domain. The measurement signal is delivered by a pulse generator. Certain characteristics of the pulse can be adjusted: the rise and fall times, the duration, the repetition rate and the amplitude. The receiver is a fast oscilloscope. The real time oscilloscope acquires the measured time response in one step, but its sensitivity is limited and it is very expensive. The sampling oscilloscope requires numerous repetitions of the measurement signal to acquire its time response, but its sensitivity is better and its price is lower than those of the real time oscilloscope. In 2009, the maximum frequency of operation for real time oscilloscope is 20GHz and 75GHz for sample oscilloscope.

Probe

Direct far-field measurements use a source antenna. The dimensions of this source antenna are limited by the distance between this antenna and the tested antenna. A large source antenna increases the measurement signal and decreases the parasitic reflections. The measured polarization is the one of the source antenna. Near-field measurements can use several types of probe: open-ended circular or rectangular metallic waveguide and electric dipole for narrow band operation (half a octave) and ridged waveguide for broad band operation (a decade).

CONCLUSION

Currently it is possible to measure all the characteristics of an antenna with a good accuracy. Far-field ranges do not have a very good accuracy, due to parasitic reflections for the outdoor ranges and because of the limited distance between the source antenna and the tested antenna for the indoor ranges. The compact range allows one to obtain a direct farfield cut in a relatively short time. The near-field techniques are the most accurate and the most convenient for global antenna radiation testing. Their main defect is the duration of the measurement which rises from the large number of necessary space points. Rapid near-field measurement systems allow one to solve this problem, but the accuracy is less good, and the frequency bandwidth is limited. Progress is necessary in this field. Research relates to the rise in frequency, with for solutions the hologram compact antenna test range and the phaseless methods. The hologram compact antenna test range must improve their accuracy while the phaseless methods must improve their reliability. Electromagnetic diagnosis of antenna must be optimized on a case-by-case basis.

REFERENCES

1. Ashkenazy, J. et al. (1985). Radiometric measurement of antenna efficiency, Electronics letters, Vol.21, N°3, January 1985, pp.111-112, ISSN 0013-5194

2. Bolomey, J. C. & Gardiol, F.E. (2001). Engineering applications of the modulated scatterer technique, Artech House Inc, ISBN 1-58053-147-4, 685 Canton Street, Norwood, MA 02062 USA

3. Bowman, R. R. (1973). Some recent developments in the characterization and measurement of hazardous electromagnetic fields, International Symposium, Warsaw, October 1973, pp217-227

4. Callen, A. L. & Parr, J. C. (1955). A new perturbation method for measuring microwave fields in free space, IEE(GB), n°102, 1955, p.836

5. Hygate, G. & Nye J. F. (1990). Measurement microwave fields directly with an optically modulated scatterer, Measurement Science and Technology's, 1990, pp.703-709, ISSN 0957-0233

6. Hirvonen, T. et al. (1997). A compact antenna test range based on a hologram, IEEE Transactions on Antennas and Propagation, Vol.AP45, N°8, August 1997, pp.1270- 1276, ISSN 0018-926X

7.   ICNIRP, (1998). Guide pour l'établissement de limites d'exposition aux champs électriques, magnétiques et électromagnétiques, Health Physics Society, Vol.74, n°4, 1998, pp.494- 522, ISSN 0017-9078

8.   Isernia, T. & Leone, G. (1994). Phaseless near-field techniques : formulation of the problem and field properties, Journal of electromagnetic waves and applications, Vol.8, N°1, 1995, pp.267-284, ISSN 0920-5071

9.   Isernia, T. & Leone, G. (1995). Numerical and experimental validation of a phaseless planar near-field technique, Journal of electromagnetic waves and applications, Vol.9, N°7, 1994, pp.871-888, ISSN 0920-5071

10. Johnson, R. C. et al. (1969). Compact range techniques and measurements, IEEE Transactions on Antennas and Propagation, Vol.AP17, N°5, September 1969, 568-576, ISSN 0018-926X

11. Johnson, R. C. et al. (1973). Determination of far-field antennas patterns from near-field measurements, Proceeding of the IEEE, Vol.61, N°12, December 1973, pp.1668-1694, ISSN 0018-9219

12. Kenefick, J. F. (1971). Ultra-precise analytics, Photogrammetry Engineering, Vol.37, 1971, pp.1167-1187, ISSN 0031-8671

13. Kummer, W. & Gillepsie, E. (1978). Antenna measurement – 1978, Proceedings of the IEEE, Vol.66, N°4, April 1978, 483-507, ISSN 0018-9219

14. Lee, J.J. et al. (1988). Near-field probe used as a diagnostic tool to locate defective elements in an array antenna, IEEE Transactions on Antennas and Propagation, Vol.AP36, n°6, June 1988, ISSN 0018-926X

15. Lemanczyk, H. (1988). Comparison of near-field range results, IEEE Transactions on Antennas and Propagation, VolAP36, n°6, June 1988, pp.845-851, ISSN 0018-926X

16. Loader, B. G. et al. (2003). An optical electric field probe for specific absorption rate measurements, The 15th International Zurich Symposium, 2003, pp.57-60

17. Newell, A. C. (1988). Error analysis techniques for planar near-field measurements, IEEE Transactions on Antennas and Propagation, VolAP36, n°6, June 1988, pp.754-768, ISSN 0018-926X

18. Picard, D. et al. (1992). Real time analyser of antenna near-field distribution, 22nd European Microwave Conference, Vol..1, pp.509-514, Espoo, Finland, 24-27 August 1992

19. Picard, D. et al. (1996). Reconstruction de la loi d›alimentation des antennes réseau à partir d›une mesure de champ proche par la méthode matricielle, Jina 1996, Novembre 1996, Nice

20. Picard, D. et al. (1998). Broadband and low interaction rapid cylindrical facility, PIERS 1998, Nantes, France, July 1998, pp.13-17

21. Picard, D. & Gattoufi, L. (1998). Diagnostic d'antennes réseau par des méthodes matricielles, Revue de l'Electricité et de l'Electronique, n°9, Octobre 1998

22. Pozar, D. & Kaufman, B. (1988). Comparison of three methods for the measurement of printed antennas efficiency, IEEE Transactions on Antennas and Propagation, Vol.AP36, N°1, January 1988, 136-139, ISSN 0018-926X

23. Rahmat Samii, Y. (1985). Microwave holography of large reflector antennas simulation algorithms, IEEE Transaction on Antennas and Propagation, Vol.AP33, N°11, November 1985, ISSN 0018-926X

24. Rahmat Samii, Y. (1988). Application of spherical near-field measurements to microwave holographic diagnosis of antennas, IEEE Transaction on Antennas and Propagation, VolAP36, n°6, June 1988, ISSN 0018-926X

25. Richmond, J. H. (1955). A modulated scattering technique for measurement of field distribution, IRE Transactions on Microwave theory and technique, Vol.3, 1955, pp.13- 17

26. Slater, D. (1991). Near-field antenna measurements, Artech House Inc, ISBN 0-89006-361-3, 685 Canton Street, Norwood, MA 02062 USA

27. Wegrowicz, L.A. & Pokuls, R.. (1991). Inverse problem approach to array diagnostics, IEEE AP-S International Symposium, Ontario, Canada, pp.1292-1295, 24-28 June 1991

28. Yaghjan, A.D. (1982). Efficient computation of antenna coupling and fields within the nearfield region, IEEE Transactions on antennas and Propagation, Vol. AP30, n°1, January 1982, pp113-127, ISSN 0018-926X

29. Yaghjian, A. (1986). An overview of near-field antenna measurements, IEEE Transactions on Antennas and Propagation, Vol.AP34, N°1, January 1986, 30-45, ISSN 0018-926X

30. Ziyyat, A. et al. (2001). Prediction of BTS antennas safety perimeter from near-field to nearfield transformation : an experimental validation, AMTA'2001 Symposium, Denver, Colorado, USA, October 2001

CITATION

CHAPTER 1

Gema Romero, Jose Ramon Díaz, Jose Maria Sabater * and Carlos Perez, Evaluation of Commercial Probes for On-Line Electrical Conductivity Measurements during Goat Gland Milking Process, doi:10.3390/s120404493.

CHAPTER 2

M. Greyson Christoforo, Eric T. Hoke, Michael D. McGehee and Eva L. Unger, Transient Response of Organo-Metal-Halide Solar Cells Analyzed by Time-Resolved Current-Voltage Measurements, doi:10.3390/photonics2041101.

CHAPTER 3

Jürgen W. Czarske, Hannes Radner, Christoph Leithold and Lars Büttner, Smart Laser Interferometer with Electrically Tunable Lenses for Flow Velocity Measurements through Disturbing Interfaces, doi:10.3390/photonics2010001.

CHAPTER 4

Yang J, Kim J, Kim W, Kim YH (2012) Measuring User Similarity Using Electric Circuit Analysis: Application to Collaborative Filtering. PLoS ONE 7(11): e49126. doi:10.1371/journal.pone.0049126.

CHAPTER 5

M. Charfeddine, M. Gassoumi, H. Mosbahi, C. Gaquiére, M. Zaidi and H. Maaref, "Electrical Characterization of Traps in AlGaN/GaN FAT-HEMT's on Silicon Substrate by C-V and DLTS Measurements," Journal of Modern Physics, Vol. 2 No. 10, 2011, pp. 1229-1234. doi: 10.4236/jmp.2011.210152.

CHAPTER 6

Alwan FM, Baharum A, Hassan GS (2013) Reliability Measurement for Mixed Mode Failures of 33/11 Kilovolt Electric Power Distribution Stations. PLoS ONE 8(8): e69716. doi:10.1371/journal.pone.0069716.

CHAPTER 7

J. Tao, C.Z. Zhao, C. Zhao, P. Taechakumput, M. Werner, S. Taylor and P. R. Chalker, Extrinsic and Intrinsic Frequency Dispersion of High-k Materials in Capacitance-Voltage Measurements, doi:10.3390/ma5061005.

CHAPTER 8

Fernando Álvarez, Fernando Garnacho, Javier Ortego and Miguel Ángel Sánchez-Urán, Application of HFCT and UHF Sensors in On-Line Partial Discharge Measurements for Insulation Diagnosis of High Voltage Equipment, doi:10.3390/s150407360.

CHAPTER 9

E. Pekşen, "Calculation of Constitutive Parameters from Electric and Magnetic Field Measurements in an Anisotropic Medium with a Triaxial Instrument," International Journal of Geosciences, Vol. 4 No. 1, 2013, pp. 58-68. doi: 10.4236/ijg.2013.41007.

CHAPTER 10

Xu X, Reinhardt JM, Hu Q, Bakall B, Tlucek PS, et al. (2012) Retinal Vessel Width Measurement at Branchings Using an Improved Electric Field TheoryBased Graph Approach. PLoS ONE 7(11): e49668. doi:10.1371/journal.pone.0049668.

CHAPTER 11

Dominique Picard (2010). Antenna Measurement, Microwave and Millimeter Wave Technologies Modern UWB antennas and equipment, Igor Mini (Ed.), ISBN: 978-953-7619-67-1, InTech, DOI: 10.5772/9023.

INDEX

A

Adaptive optics (AO) 50, 60
Animal Production 3
Atomic force microscopy (AFM) 106, 144
Atomic layer deposition (ALD) 106, 143

B

Bipartite graph 68, 70, 71

C

Capacitance equivalent thickness (CET) 108, 113, 146, 151
Capacitance-Voltage 89, 90, 91, 92
Central Retinal Vein Equivalent (CRVE) 256
Collaborative filtering (CF) 66
Complementary-metal-oxide-semiconductor (CMOS) 102, 140
Connection satellite 90
Coordinated universal time (CUT) 195
Current-deep level transient spectroscopy (CDLTS) 90

D

Deep level transient spectroscopy (DLTS) 90
Deep Level Transient Spectroscopy (DLTS) 89
Doppler frequency 51

E

Electrical conductivity (EC) 1
Electrical network 68, 70
Electric circuit 65, 68, 69, 71, 73, 81, 84
Electric circuit analysis (ECA) 84
Electric-circuit analysis (ECA) 66
Electric line of force (ELF) 246
Electron trap 89, 94, 95
Equivalent oxide thickness (EOT) 102, 140
Evidence corroborates 31

F

Ferroelectric polarizability 31
Field distribution 31, 37, 39
Fluid mechanics 49, 50, 61
Fraunhofer region 262

G

Galvanometer mirrors (GM) 55
Global positioning system receiver
    (GPS) 195

H

Hazard and critical control points
    (HCCP) 1
High electron mobility transistors
    (HEMT's) 89
High frequency current transformers
    (HFCT) 178
High voltage (HV) 177

I

International Technology Roadmap for
    Semiconductors (ITRS) 102, 140

K

Kirchhoff's current law (KCL) 72
Kirchhoff's voltage law (KVL) 72
Kohlrausch-Williams-Watts (KWW)
    101, 103, 126, 139, 141, 163

L

Laser Doppler velocimeter (LDV) 55
Laser Doppler velocimetry (LDV) 50

M

Metal-oxide-semiconductor (MOS) 101,
    102, 104, 139, 140, 141
Methylammonium iodide (MAI) 32
Metropolis–Hastings random walk
    (MHRW) 74
Micro-electro-mechanic-system
    (MEMS) 53

N

Negative electrodes 11

O

Optical coherence tomography (OCT)
    50
Optical distortions 50, 52, 56, 61
Organo-metal-halide perovskite (OMHP)
    29

P

Partial discharge (PD) 177
Pattern integration method 264
Physical phenomena 30, 38
Potential values 66
Principal component analysis (PCA)
    244
Pulse per second signal (PPS) 195

R

Radio frequency current transducer
    (RFCT) 180
Radiolocalisation systems 261
Reverse scans 30, 37
Root Mean Square (RMS) 117, 155

S

Signal to noise ratio (SNR) 182, 190
Signal-to-noise ratio (SNR) 52
Static experiments 8
Static process 4

T

Tanimoto coefficient (TC) 66
Tunable lenses (TL) 55
Tunable lens (TL) 57

V

Velocity measurements 49, 50, 56, 61,
    62

W

Wavelet transform (WT) 193

X

X-ray diffraction (XRD) 106, 144